Emerging Contaminants: Anticipating Developments

Emerging Contaminants: Anticipating Developments

Kathleen Sellers
Denice Nelson
Nadine Weinberg

CRC Press
Taylor & Francis Group
Boca Raton London New York

CRC Press is an imprint of the
Taylor & Francis Group, an **informa** business

CRC Press

Taylor & Francis Group

6000 Broken Sound Parkway NW, Suite 300

Boca Raton, FL 33487-2742

© 2020 by Taylor & Francis Group, LLC

CRC Press is an imprint of Taylor & Francis Group, an Informa business

No claim to original U.S. Government works

Printed on acid-free paper

International Standard Book Number-13: 978-0-367-20979-7 (Hardback)

**Visit the Taylor & Francis Web site at
http://www.taylorandfrancis.com**

**and the CRC Press Web site at
http://www.crcpress.com**

Contents

Preface

The emergence of new concerns about a contaminant can alert the public, too late, to exposure and consequent risk that have already occurred; spur new legislation or regulations with far-reaching consequences; and upend a company's plans for site remediation or use of a product. This book proposes a bold idea: we can learn from the emergence of "new" contaminants in the past to gain insights into the contaminants that may emerge as new concerns in the future. Such forewarning offers the opportunity to protect human health and the environment. It can also enable a company to anticipate and manage its business risks.

The analyses described in this book originate in multiple disciplines: the science of toxicology; environmental law and regulation; the field of product stewardship; and the social science which explains why ideas take hold. The results of the analyses in this book support a step-by-step method to assess the potential for a contaminant to emerge, and a framework to apply those conclusions to managing site liabilities. The authors' hope is that this multidisciplinary approach contributes materially to the appropriate recognition of and response to environmental contamination.

Acknowledgements

The authors gratefully acknowledge the following contributors to the data analyses reflected in this book: Jennifer Byrd, PE of ERM, and Alexis Palmer of New York University. Our thanks also go to the team at CRC Press for their support in bringing this book to publication.

Thanks go, too, to our colleagues and clients who engaged in thoughtful conversations that helped us to shape and focus our ideas about this subject. In particular, we deeply appreciate the discussions with Dr. Maureen Leahy of ERM that helped to shape the content of Section 4.2.2.

We would also like to acknowledge the inspiration of Rachel Carson. While some of her work remains controversial to this day, Ms. Carson started an important conversation about balancing the risks and rewards of chemicals in our lives. We hope that this book contributes materially to that conversation.

About the Authors

Kathleen Sellers, PE, leads multidisciplinary consulting teams that help global companies to meet their business goals through effective product stewardship and sustainability. This is her fifth book with CRC Press. From her first book, *Fundamentals of Hazardous Waste Site Remediation* (1996), through her penultimate publication, *Product Stewardship, Life Cycle Analysis and the Environment* (2015), she has explored the challenges of recognizing and responding appropriately to emerging environmental issues relating to chemicals and nanomaterials. Kate has used the content of her books to generate broader conversations through teaching courses at Tufts University, Indiana University, and Harvard University Extension School. She is a founding Board member of the Product Stewardship Society and served as President from 2017 to 2019. Kate earned a Bachelor of Science degree in Chemistry from Indiana University and a Master of Science degree in Environmental Engineering from the University of Massachusetts.

Denice Nelson, PhD, PE, a partner at ERM, specializes in remedial strategy development. In the course of her work, she has developed site characterization and remediation approaches for a variety of contaminants, both conventional and emerging. She brings to her work a keen appreciation for the technical, regulatory, and public perception factors that shape decisions about emerging contaminants in the environment. She holds a BCE and MS in Civil Engineering and a PhD in Environmental Engineering from the University of Minnesota.

Nadine Weinberg has been working in the field of environmental risk assessment for 30 years. Her work on this book reflects various perspectives on risk assessment. She worked for the US Environmental Protection Agency to support chemical regulation related to the Superfund program and air quality. Over the last 25 years, Nadine has been primarily focused on working with commercial and industrial companies to understand potential human and ecological risks from chemical exposures in the environment. Nadine earned a Bachelor of Science degree in Natural Resources from Cornell University and a Master of Environmental Management degree from Duke University.

1 Introduction

News headlines quote activists outraged about contamination near their homes and in their drinking water. Scientific papers about the contaminant proliferate. Regulators react by developing new cleanup goals that change site investigation and cleanup requirements at the cost of millions of dollars. Products may be pulled from store shelves. Each time, this scenario catches us by surprise, and yet, each time, the pattern feels strikingly familiar.

Perhaps this scenario need not come as a surprise. This book examines the factors that have led "new" environmental contaminants to emerge and combines lessons learned from that history with data from the field of product stewardship to anticipate potential new developments. It provides readers a framework for anticipating the emergence of new contaminants so that the risks – whether to human health and the environment or to a business – can be anticipated and appropriately managed.

1.1 APPLYING LESSONS FROM THE PAST TO ANTICIPATE THE FUTURE

The challenge of recognizing and responding to emerging contaminants is nothing new. Consider just two examples. Sanskrit records circa 2000 BC advised that water could be purified by charcoal filtration and exposure to sunlight, a precursor to today's treatment by activated carbon and ultraviolet oxidation [1]. Waste from coke ovens contaminated many water supplies early in the last century. In 1917, scientists who had observed pollution by coke oven waste published a study of the toxicity of those wastes to fish, noting that, "[t]he products of destructive distillation of coal include an innumerable series of substances ... their injurious effect upon fishes and other life of streams generally is itself sufficient to justify the prohibition of pollution by [discharge to surface water]" [2]. In the 1920s, engineers in England devised ways to treat drinking water to remove the objectionable taste and odor of coke oven wastes [1]. These stories from the past illustrate two of the key factors that cause a contaminant to emerge: exposure is found to occur, perhaps initially by recognizing that drinking water is contaminated, and exposure to the contaminant causes an undesirable effect.

Emergence of a contaminant in current times can reflect additional factors [3], as illustrated in Figure 1.1.* Recognition of a new issue sometimes starts with "fringe" observations in a particular location or by a scientist working in a related field of study. Some issues may never draw more attention or may linger at the fringes for decades. The history of perfluoroalkyl and polyfluoroalkyl substances (PFAS), a broad class of chemicals estimated to comprise over 3,000 different compounds [4], neatly illustrates this point as shown in

1

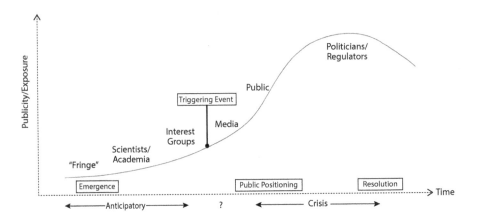

FIGURE 1.1   Life Cycle of an Issue

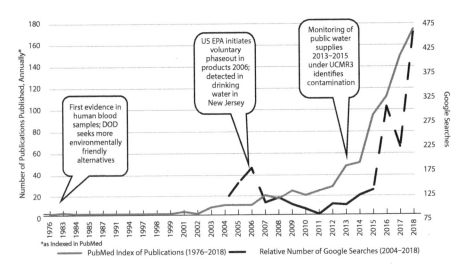

FIGURE 1.2   Growing Interest in PFAS

Figure 1.2 [5,6]. Now widely known as a contaminant that emerged circa 2015, PFAS first sparked concerns almost 40 years before that [7]. In the mid-1970s scientists had measured "fluorocarbon carboxylic acids" in samples of human blood in two states in the United States [8] and were questioning the source. In an unrelated effort at about that time, officials at the Department of Defense commissioned a study of firefighting foams as alternatives to PFAS-based foams on the basis that "improvements are desired in the environmental area, i.e., development of compositions that have a reduced impact on the environment without loss of fire suppression activities" [9].

Three decades later studies found perfluorooctanoic sulfonate (PFOS) and perfluorooctanoic acid (PFOA), two PFAS chemicals, in public water supplies in New Jersey. While the effects of such exposure were unclear, US EPA was concerned enough to add PFAS to the list of contaminants to be monitored in public water supplies [10]. The results of that nationwide monitoring program showed PFAS contamination in drinking water supplies across the United States [11] and led to public alarm, scientific research, and ultimately regulatory action.

The initial findings about a particular contaminant may draw the attention of research scientists. Their work may produce more evidence of exposure or harm and raise the level of concern. Publications can generate more funding to generate more data; cynics may agree with the academics who wrote in 2014 "[e]merging contaminants have now become a fashionable and trendy research venue" [12]. Figure 1.2 shows how the number of scientific publications on PFAS[**] grew through the mid-2000s and escalated rapidly after 2010 as PFAS drew increasing attention.

As the scientific evidence of possible risks begins to accumulate, it may catch the attention of activists or the media. Concerns can then escalate rapidly. Figure 1.1 shows this pattern in a general sense. Figure 1.2 illustrates the phenomenon for PFAS based on the relative number of Google searches as evidence of public interest. As concerns about a contaminant develop into a broader awareness with increasing public attention, regulators often begin to take action. The attention paid to PFAS, as illustrated in Figure 1.2, has led to various regulatory responses intended to minimize the risk from PFAS [13].

Finally, in many cases, the once-burning issue becomes more routine. But, by that time, damage has been done. As in the case of PFAS (although PFAS issues are not yet routine as of this writing) people and organisms in the environment have been exposed to the contamination; millions of dollars in unanticipated costs have been spent.

What if we could anticipate the emergence of the next contaminants and mitigate those effects? While no one can answer that question with absolute certainty, the emergence of a contaminant of concern is hardly random. We can look backward in time to determine the causative factors and, using publicly available data, look ahead to what may – and may not – emerge as a concern in the future.

The remainder of this book combines insights from past examples of emerging contaminants with new data drawn from multiple disciplines in order to anticipate the emergence of new contaminants. While the authors used data largely from the United States and Europe, we anticipate that the conclusions will be broadly relevant to conditions in other parts of the world. And while this book focuses on contaminants at hazardous waste sites, the analyses presented here also pertain to environmental contamination more generally.

Chapter 2 lays the groundwork by exploring why we investigate and remediate the contaminants that we do. It probes the history of environmental regulations, the evolution of analytical techniques for environmental samples, and the way in which we have understood hazard and therefore risk, to examine the basis for current practices and the gaps that can allow new contaminants to emerge.

In Chapter 3 we'll explore the factors that propel the emergence of a contaminant into broad consciousness: a new understanding of the extent of exposure, insights into possible hazards, and the outrage that leads to calls for action. This chapter describes the factors that can result in the emergence of a "new" contaminant, illustrating each factor with examples. Chapter 4 builds on the insights described in Chapters 2 and 3 and provides the reader tools to understand and even anticipate the emergence of new contaminants.

1.2   DATA USED IN ANALYSES

Predictions without factual basis lack credibility. The analyses in Chapters 2 and 3 therefore reflect data that describe chemicals in commerce, regulatory constructs, and the nature of certain contaminants. Table 1.1 lists the data sets used throughout this book to extrapolate from past lessons in order to antici-pate contaminants that may emerge as concerns in the future. These data sources are each described in more detail below.

TABLE 1.1

Sources of Data

Factor	List	Why the list is important
Regulated Chemicals	CERCLA List	Hazardous substances investigated under federal law governing haz-ardous waste site investigation and remediation in the United States
	TSCA Work Plan	Chemicals in commerce in the United States identified for risk assess-ment based on potential for hazard and exposure to create unreason-able risk; many of these chemicals have not undergone formal risk assessment previously
Potential for Exposure	HPV	Chemicals manufactured or imported at 1 million pounds (453,592 kilograms) per year or more in the United States
	TRI	Toxic chemicals emitted from industrial facilities and federal facilities in the United States
	UCMR	Unregulated drinking water contaminants designated for monitoring by the US EPA, which may produce data that show evidence of con-tamination and cause a "new" contaminant to emerge into recognition
Potential for Hazard	IRIS	US EPA's compilation of toxicity data, largely amassed in the 1980s, used as a basis for determining cleanup goals
	PBT	Chemical compounds in commerce that are persistent, bioaccumula-tive, and toxic, and therefore present hazards of especial concern
	PMOC	Persistent, mobile organic compounds have the potential to pass through conventional water treatment processes and become environ-mental contaminants
	SVHC	Substances of Very High Concern, designated by regulators in the European Union as a result of their significant hazards
	SIN List	Compounds on a list maintained by a nongovernmental organization which have the characteristics of SVHCs but are not yet regulated

1.2.1 REGULATED CHEMICALS

This book references two main sources of regulatory information about known contaminants and the potential for new contaminants to emerge. The first, abbreviated in this book as CERCLA List [14], comprises the list of approximately 1,000 hazardous substances designated under the US Comprehensive Environmental Response, Compensation, and Liability Act of 1980 (CERCLA), commonly known as Superfund. (Radionuclides listed under CERCLA and CERCLA hazardous substance chemical categories were omitted from the analyses in this book.) Regulations at 40 CFR 302.4 designate hazardous substances under the Superfund, Emergency Planning, and Community Right-to-Know Programs. This list represents the maximum number of contaminants typically targeted for investigation and remediation at hazardous waste sites. In most cases, and as discussed further in Section 2.3, most site investigations focus on far fewer chemicals.

The Toxic Substances Control Act (TSCA) Work Plan identifies the chemicals in commerce in the United States that the US EPA has identified as needing further assessment. The 90 listed chemicals have been manufactured in or imported into the United States for decades but not yet, according to current scientific practice, assessed for hazard or risk under these use scenarios. The US EPA has indicated that [15]:

> The screening process for identifying these chemicals is based on a combination of hazard, exposure (including via uses [sic]), and persistence and bioaccumulation characteristics.... The Agency continues to use this process, which focuses on chemicals that meet one or more of the following factors:

- Potential concern for children's health (for example, because of reproductive or developmental effects)
- Neurotoxic effects
- Persistent, bioaccumulative and toxic
- Probable or known carcinogens
- Used in children's products or in products to which children may be highly exposed
- Detected in biomonitoring programs.

1.2.2 POTENTIAL FOR EXPOSURE

Contaminants sometimes emerge as a concern when new data show that widespread exposure is occurring. The analyses in this book reference several databases that represent the potential for exposure. Chief among these is the list of High Production Volume (HPV) chemicals, i.e., the approximately 4,300 chemicals manufactured or imported in more than 1 million pounds (approximately 450,000 kilograms) per year in the United States based on Chemical Data Reporting as recorded in the US EPA's ChemView database in 2019 [16]. (In the early days of US EPA's HPV program circa 1998, the Agency identified approximately 2,800 HPV chemicals [17].) The simple fact that a chemical is manufactured or imported at a high volume does not mean that it will become a significant environmental contaminant. However, the HPV list does provide a crude indication of the chemicals that have the potential to be released into the environment in the greatest quantities. Whether they are or not depends upon the potential for discharge during manufacture, use or disposal, and the possibility that an accident could release the substance to the environment.

A second database, known as the Toxics Release Inventory (TRI) [18], provides information on the chemicals released to the environment under permit in the US during routine manufacturing operations. The US EPA describes this reporting program as follows [19]:

> U.S. facilities in different industry sectors must report annually how much of each chemical is released to the environment and/or managed through recycling, energy recovery and treatment. (A "release" of a chemical means that it is emitted to the air or water, or placed in some type of land disposal.) ...
>
> In general, chemicals covered by the TRI Program are those that cause:
>
> • Cancer or other chronic human health effects
> • Significant adverse acute human health effects
> • Significant adverse environmental effects.There are currently over 650 chemicals covered by the TRI Program. Facilities that manufacture, process or otherwise use these chemicals in amounts above established levels must submit annual TRI reports on each chemical. The TRI chemical list doesn't include all toxic chemicals used in the U.S.

The TRI list includes 695 individually listed chemicals and 33 chemical categories.

Under the Unregulated Contaminant Monitoring Rule (UCMR), the US EPA has the authority to monitor drinking water for contaminants that do not have health-based standards set under the Safe Drinking Water Act and that the Agency suspects to be present in drinking water. Data from this program have led contaminants to "emerge" as concerns in the past, and consequently the ever-evolving monitoring lists [20] under UCMR should be considered in assessing the potential that regulators may find new evidence of widespread exposure to a previously unrecognized contaminant.

1.2.3 POTENTIAL FOR HAZARD

Fresh concerns over environmental contamination can emerge when new scientific data show that exposure to previously unrecognized hazards can result in significant risks. This book references five lists that catalog potential chemical hazards. Each list or category of substances evolves over time with the addition of new chemicals; the lists are current as of the references cited.

• The Integrated Risk Information System (IRIS) [21] contains dose–response factors representing US EPA's scientific positions on the cancer and noncancer human health effects that could result from chronic oral, dermal, or inhalation exposure to chemicals in the environment. This database is a primary reference for characterizing the hazards of 568 contaminants.

- PBT chemicals, known to be persistent, bioaccumulative, and toxic, do not attenuate naturally, may be difficult to remediate, and are likely to have relatively low cleanup goals as a result of their toxicity. Most people working in the environmental field are familiar with polychlorinated biphenyls, probably the best-known PBT chemicals. But few may be aware that many lesser-known compounds are PBT. Strempel et al. [22] screened approximately 95,000 chemicals in commerce in the European Union based on estimated P, B, and T properties to determine which substances might be classified as PBT. The resulting list of approximately 2,800 chemicals was used in analyses described in this book to assess the potential for emerging environmental contaminants to be identified based on their persistent, bioaccumulative, or toxic nature. These comparisons reflect the implicit assumption that chemicals in commerce in the European Union are likely to be or have been in commerce in the United States. (See Section 1.2.4.)
- Persistent, mobile organic compounds (PMOC) or persistent, mobile and toxic (PMT) compounds are loosely defined class(es) of contaminants of recent concern. Regulators and research teams in Europe have begun to assess the occurrence of such chemicals in the environment, and the need to regulate them. The data comparisons in this book are based on a study [23] that identified 2,167 unique substances (including organic and pseudo-organic chemical compounds) in commerce which may be of concern in this category.
- Substances of Very High Concern, or SVHC, are designated under Article 57 of the European Union regulation, Registration, Evaluation, Authorisation and Restriction of Chemicals (REACH). The European Chemicals Agency (ECHA) maintains a database of substances designated as SVHC under REACH, numbering 197 substances as of May 2019 [24]. These chemicals meet one or more of the following criteria:

  - Substances meeting the criteria for classification as carcinogenic, mutagenic, or toxic for reproduction (CMR) category 1A or 1B in accordance with the Classification Labelling and Packaging Regulation.
  - Substances which are persistent, bioaccumulative, and toxic (PBT) or very persistent and very bioaccumulative (vPvB) according to REACH Annex XIII.
  - Substances on a case-by-case basis that cause an equivalent level of concern as CMR or PBT/vPvB substances.
- The Substitute it Now (SIN) List [25] comprises 919 chemicals likely to be restricted or banned in the EU based on analyses by the International Chemical Secretariat (known as ChemSec), which is a nonprofit organization supported in part by the Swedish Government. ChemSec uses the SVHC criteria established under REACH to identify substances that are not yet regulated but may be in the future.

1.2.4 Putting the Data into Context

Consider the following simple comparisons based on the information cited above. Of the 40,655 chemicals in commerce in the United States based on the TSCA Inventory [26], 4,308 are produced at over 1 million pounds (approximately 450,000 kilograms) per year. Not all of these chemicals are particularly hazardous. But regulators, scientists, and nongovernmental organizations have identified thousands of chemicals in commerce that are particularly persistent, bioaccumulative, toxic, and/or mobile and thus of special concern as potential contaminants. Many of these substances are manufactured in large quantities. Yet site investigation and remediation commonly focus on approximately 1,000 or fewer hazardous substances based on decades-old regulations; of these chemical compounds, the US EPA has developed scientific positions on human health effects for just 568 compounds.

These simple comparisons show the potential for new contaminants to emerge as concerns. One might conclude that hundreds of "new" contaminants could soon warrant action. But we must not rush to judgment. The potential for any one chemical compound to present the sort of widespread and significant risk that calls for action depends upon a complex mix of variables only partially captured by the data analyses in this book: the extent of releases, permitted and otherwise; fate and transport in the environment and potential for exposure; ability to detect a contaminant in the environment; hazards to human health and the ecosystem by the chemical and its degradation products; and the balance of risks and rewards associated with any action, including the manufacture and use of a chemical. Further, the emergence of a "new" contaminant often depends upon public outrage and the degree to which that outrage spreads.

Throughout this book, the authors have chosen to describe the results of data comparisons in general terms by providing the numerical results – numbers of chemicals or mass per year, for example – and then illustrating key points with specific examples. We look to better quantify the "X" factor of public outrage with several data-driven comparisons looking backwards in time until present day. We also map out in Chapter 4 a logical method to assess the potential for a "new" contaminant to emerge. We have chosen, for the most part, not to include lists of target chemicals deduced from the comparisons in this book because of the potential that such lists could be taken out of context. Without reasoned context, such lists can easily provoke outrage that can distract from the science-based assessment and resolution of truly significant risks.

The context for the analyses in this book includes limitations on their accuracy. Readers should be aware of three limitations to the data assessments described herein. First, the lists described in Section 1.2 were compared by their Chemical Abstract Services (CAS) numbers. Some lists (e.g., the TSCA Work Plan) describe a few chemical compounds by name or category and not by CAS number. As a result, the "counts" of chemicals common to two or more lists may not be precisely accurate. The authors believe that this inaccuracy does not detract from the overall validity of our conclusions.

The second limitation pertains to comparisons of specific hazards, as identified in the European Union, to chemicals in commerce in the United States. Researchers and regulators in the European Union have done much of the work to identify particularly hazardous chemicals, e.g., as SVHC, PBT, or PMT/PMOC. They have focused on chemicals in commerce in the European Union. As of June 2019, 22,213 unique substances are manufactured in or imported into the European Union in quantities greater than one metric ton per year [27]. When researchers developed the PBT and PMOC/PMT lists cited in this book they based their analyses on the chemicals in commerce in the European Union at the time of their research. Thus, the PBT and PMOC/PMT lists used in comparisons in this book may not precisely represent the number of PBT and PMOC/PMT chemicals in commerce in the United States. It is also possible that PBT or PMOC/PMT chemicals in commerce in the EU are not in commerce in the United States.

Finally, each list described in this section and used in comparisons throughout the book represents conditions at the point in time when regulators or scientific teams compiled the list. Manufacturing conditions change, scientific information on chemical hazards grows, and so each list may evolve over time.

The evaluations in this book and the resulting conclusions largely focus on the investigation and remediation of contaminated sites. However, contamination is not limited to uncontrolled hazardous waste sites. As illustrated by examples in this book, a contaminant may first be discovered in drinking water supplies and then, upon investigation, in groundwater plumes from waste sites. Alternatively some environmental concerns relate to conditions in surface water or soil generally associated with the manufacture and use of chemicals and not with a specific release at a particular site. In consequence, while the authors write largely from the perspective of contaminated site investigation and remediation, we have relied on information sources and draw conclusions regarding potential risks that can apply to environmental contamination more broadly.

1.3 MANAGING RISKS

The emergence of a "new" environmental contaminant raises awareness that human and ecological systems may have been exposed to chemicals. The effects may not be known. But harm may have already occurred. For a company managing its environmental liabilities, the emergence of a contaminant can also present business risks. Understanding the potential for new contaminants to emerge is crucial to mitigating risks to human health and the environment, and to business. This book will give the reader tools to understand and even anticipate the emergence of new contaminants.

We offer these tools with caution and with hope. Expert predictions are notoriously fallible. One study has found that the best predictions emerge from teams that "draw from an eclectic array of traditions, and accept ambiguity and contradiction" rather than from tightly focused specialists [28]. In that spirit the authors drew the analyses described in this book from many disciplines: the science of toxicology; environmental law and regulation; the field of product stewardship; and the social science which explains why ideas take hold. Our hope is that this multidisciplinary approach, despite ambiguities and contradictions in the analytical results, contributes materially to the appropriate recognition of and response to environmental contamination.

NOTES

* Figure 1.1 is adapted from Figure 3.3 – The general life cycle of an issue, Iannuzzi, A., 2018. *Greener Products: The Making and Marketing of Sustainable Brands.* Second Edition. CRC Press, and is used with permission.
** Figure 1.2 indicates the number of publications each year according to PubMed. This database, which is maintained by the US National Library of Medicine National Institutes of Health, contains 29 million citations for biomedical literature from MEDLINE, life science journals, and online books. Thus, the number of papers published each year provides an indication of the level of research into exposure to and hazards of a chemical substance.

REFERENCES

1. Baker, M.N., 1948. *The Quest for Pure Water.* New York: The American Water Works Association, pp. 449–453.
2. Shelford, V.E., 1917. An experimental study of the effects of gas waste upon fishes with especial reference to stream pollution. *Illinois Natural History Survey Bulletin*, 11(4), pp. 381–412.
3. Iannuzzi, A., 2018. *Greener Products: The Making and Marketing of Sustainable Brands*, 2nd ed. Boca Raton, FL: CRC Press. Chapter 3.
4. Interstate Technology Regulatory Council, 2017. Per- and Polyfluoroalkyl Substances (PFAS) fact sheets: Introduction.
5. US National Library of Medicine National Institutes of Health. PubMed. Results of Key Word Search for "PFAS". Available at: www.ncbi.nlm.nih.gov/pubmed/?term=Perchlorate (accessed June 10, 2019).
6. Google Trends. Keyword Search for "PFOS + PFOA + PFAS". Available at: https://trends.google.com/trends/explore?date=all&geo=US&q=PFOS%20PFAS%20PFOA (accessed June 10, 2019).
7. Hayes, J., & Faber, S., 2019. Mapping PFAS Chemical Contamination at 106 US Military Sites: the Pentagon's 50-year History with PFAS Chemicals. Available at: www.ewg.org/research/pfas-chemicals-contaminate-least-110-us-military-sites/pentagon-s-50-year-history-pfas (accessed April 14, 2019).
8. Crawford, G.H., 1975. Fluorocarbons in Human Blood Plasma [Memorandum, 3M]. August 20. Available at: www.documentcloud.org/documents/4558283-Dr-Guy-Phone-Call-to-3M.html (accessed April 14, 2019).
9. Peterson, H.B., 1976. R&D Final Report on DOD-AGFSP.S-76-10 (MIPR FY 7615-76-05063) – Improved Environmental Impact Properties for AFFF Materials. October. Available at: https://assets.documentcloud.org/documents/4344668/122.pdf (accessed April 14, 2019).
10. US EPA, 2009. Summary of Nominations for the Third Contaminant Candidate List. Appendix 1. Available at: www.epa.gov/sites/production/files/2014-05/documents/nomination_summary083109_508_v3.pdf (accessed June 11, 2019).

11. US EPA, 2017. The Third Unregulated Contaminant Monitoring Rule (UCMR 3): Data Summary, January 2017. Available at: www.epa.gov/sites/production/files/ 2017-02/documents/ucmr3-data-summary-january-2017.pdf (accessed April 14, 2019).
12. Sauvé, S., & Desrosiers, M., 2014. A review of what is an emerging contaminant. *Chemistry Central Journal*, 8(1), 15.
13. ITRC, 2018. Regulations, Guidance, and Advisories for Per- and Polyfluoroalkyl Substances (PFAS). Available at: https://pfas-1.itrcweb.org/wp-content/uploads/ 2018/01/pfas_fact_sheet_regulations__1_4_18.pdf (accessed June 10, 2019).
14. US EPA, 2015. Consolidated List of Lists under EPCRA/CERCLA/CAA §112(r) (March 2015 Version). Available at: www.epa.gov/epcra/consolidated-list-lists-under -epcracerclacaa-ss112r-march-2015-version (accessed January 2019).
15. US EPA, 2014. TSCA Work Plan for Chemical Assessments: 2014 Update. Environmental Protection Agency, Office of Pollution Prevention and Toxics. October 2014. Available at: www.epa.gov/assessing-and-managing-chemicals-under-tsca/tsca-work-plan-chemical-assessments-2014-update (accessed January 11, 2019).
16. US EPA, 2017. HPV Chemicals, exported from ChemView database. Available at: https://java.epa.gov/chemview# (accessed May 10, 2019).
17. US EPA, 2012. High Production Volume (HPV) Challenge – Basic Information. Web page last updated April 12, 2012. Available at: www.epa.gov/hpv/pubs/general/ basicinfo.htm (accessed July 12, 2013).
18. US EPA, 2019. TRI-Listed Chemicals, Reporting Year 2018. Available at: www. epa.gov/toxics-release-inventory-tri-program/tri-listed-chemicals (accessed May 2019).
19. US EPA, 2019. Learn about the Toxics Release Inventory. Web page last updated May 19 2019. Available at: https://www.epa.gov/toxics-release-inventory-tri-program /learn-about-toxics-release-inventory (accessed May 30, 2019).
20. US EPA, 2018. Drinking Water Contaminant Candidate List (CCL) and Regulatory Determination. Web page last updated October 9, 2018. Available at: www.epa.gov /ccl (accessed May 30, 2019).
21. US EPA, 2019. IRIS Assessments. Available at: https://cfpub.epa.gov/ncea/iris_ drafts/atoz.cfm?list_type=alpha (accessed January 11, 2019).
22. Strempel, S., Scheringer, M., Ng, C.A., & Hungerbühler, K., 2012. Screening for PBT chemicals among the "existing" and "new" chemicals of the EU. *Environmental Science & Technology*, 46(11), pp. 5680–5687.
23. Arp, H.P.H., Brown, T.N., Berger, U., & Hale, S.E., 2017. Ranking REACH registered neutral, ionizable and ionic organic chemicals based on their aquatic persistency and mobility. *Environmental Science: Processes & Impacts*, 19(7), pp. 939–955.
24. ECHA, 2019. Candidate List of Substances of Very High Concern for Authorisation (published in accordance with Article 59(10) of the REACH Regulation). Available at: https://echa.europa.eu/candidate-list-table (accessed January 11, 2019).
25. ChemSec, 2019. SIN List. Last updated January 2019. Available at: https://sinlist. chemsec.org/#filters (accessed January 11, 2019).
26. US EPA, 2019. EPA Releases First Major Update to Chemicals List in 40 Years. Press release February 19, 2019. Available at: www.epa.gov/newsreleases/epa-releases-first-major-update-chemicals-list-40-years (accessed May 1, 2019).
27. ECHA, 2019. Registered Substances. Web site last updated June 5, 2019. Available at: https://echa.europa.eu/information-on-chemicals/registered-substances;jsessio nid=C8B11FF15E219166EFD8031625A61A03.live2 (accessed June 5, 2019).
28. Tetlock, P.E. as Cited in Epstein, D., 2019. The peculiar blindness of experts. *The Atlantic*, 20–22. June 2019.

2 We Find What We Look For

This chapter describes the basis for accepted practice in site investigation and remediation and the gaps in that practice from which new contaminants can emerge. The past six decades have shown that our ability to identify and understand risks and the resulting determination to regulate exposures have driven how we define environmental contaminants of concern. In this section of the book, we'll explore the evolution of the factors that have influenced how we have defined environmental contamination, including:

- Environmental regulations
- Analytical techniques
- Understanding of hazard and therefore risk.

2.1 LEARNING FROM THE PAST

We'll begin with a story that illustrates the key themes of this chapter. An emerging contaminant of the late 1950s and early 1960s, orthonitrochlorobenzene (ONCB), illustrates how new understandings of hazard and risk cause a contaminant to emerge.[1] As discussed throughout this book, such evolving views of hazard and risk often occur within the context of environmental regulations and in conjunction with scientific developments that allow chemists to detect a previously unknown contaminant in environmental samples.

Few environmental laws or regulations existed in the United States in the 1950s. One of those laws, the Federal Water Pollution Control Act (FWPCA) of 1948, largely referred to planning and coordination activities intended to limit pollution in interstate waterways. Not until Congress amended the FWPCA in 1956 did the Federal government have the authority to regulate the *quality* of water that flowed across state lines. Just as important, the 1956 amendments provided for Federal grants to construct municipal wastewater treatment facilities.

In order to establish trends in contamination levels in major river systems – and ideally to document progress in cleaning up streams as a result of wastewater treatment plants funded though the FWPCA – the US Public Health Service established a Water Quality Network. The initial network provided water quality data at 50 surface-water monitoring stations around the country. It is no exaggeration to say that the establishment of this network catalyzed the development of environmental analytical chemistry and toxicology in the United States.

The scientists working on this project lacked most of the tools we now take for granted. Collecting and analyzing samples required extraordinary effort. The US Public Health Service devised means to collect samples at each station on the order of 3,000 to 4,000 gallons (approximately 11,400 to 15,100 liters) of water each and then passed each of those samples through a carbon filter to isolate organic contaminants. Analytical chemists separated the organic residues into fractions based on the solubility of chemical families in various solvents, then used infrared spectroscopy and the melting points of solids to identify individual compounds [1,2].

This painstaking work produced startling results. Researchers found ONCB in water samples collected over 900 miles (approximately 1,500 kilometers) of the Mississippi River in 1958. They traced the compound to a single apparent source near the city of St. Louis, Missouri, a discharge from a factory that manufactured paranitrochlorobenzene and produced ONCB as an "unusable byproduct" [3]. (This source was apparently a facility located in Sauget, Illinois [4], across the river from St. Louis.)

Samples from the monitoring station approximately 100 miles downstream of the apparent source, at Cape Girardeau, Missouri, typically contained the highest concentrations of ONCB detected in the US Public Health Service study. All of the samples collected each month from this station during 1958 contained ONCB, at an average concentration of 21 parts per billion (ppb). Researchers from the US Public Health Service made a conservative estimate that 22,600 pounds (approximately 10,300 kg) of ONCB flowed past this monitoring point every 24 hours [3].

The data from the monitoring station near Cape Girardeau pointed to another problem. The city of Cape Girardeau obtained its drinking water from the Mississippi River and the raw water intake contained ONCB. When the US Public Health Service analyzed samples of treated water, the data showed that drinking water treatment did not reduce the concentration of ONCB [3]. As a result the residents of Cape Girardeau ingested ONCB with their drinking water; a sample of treated water collected during one week in 1958 contained 30 ppb ONCB. These data were among the first to quantify the exposure to an anthropogenic chemical from an environmental source in the United States, or, in other words, to identify exposure to an emerging contaminant.

The scientists from the US Public Health Service also lacked the risk assessment tools we rely on now to understand the implications of exposure to chemical substances. The toxicity of ONCB had not been widely studied. Observations of workers manufacturing chloronitrobenzenes in 1923 indicated the effects of acute exposure [5]:

Workers generally required hospitalization after about 3 days of exposure. Symptoms included a slate gray appearance, headache, and dyspnea on exertion. Blood serum was often a port wine color, and erythrocytes were large and occasionally deformed. Workers received treatments of atropine, ether, pituitary extract, and a coffee enema. Despite this, all recovered over a period of several weeks.

Based on the toxicological data available in the late 1950s, scientists at the US Public Health Service estimated that the probable lethal dose of ONCB to a human was between 5 and 50 milligrams per kilogram (mg/kg) of body weight, or "between 7 drops and 1 teaspoonful for [a] 150 lb.-man. A clinically significant illness would be expected to result from 1/10 the probable lethal dose" [3]. They further calculated that the estimated daily dose would allow for a 467-fold factor of safety but concluded that, given the uncertainties, "it does not seem reasonable to have such an additive in drinking water in any concentration when it is possible to eliminate it at the source."

The results of subsequent studies refined scientists' understanding of the possible hazards. Based on the possible carcinogenicity of ONCB, the US EPA has published a non-enforceable limit on ONCB in drinking water of 0.23 ppb or micrograms per liter [6]. This guideline is two orders of magnitude below the concentration detected in drinking water in 1958.

The emergence of ONCB as a public health concern had several effects in the short and long term. An addendum to the US Public Health Service report on the study of Cape Girardeau drinking water written in 1959 notes that "[o]nce this chemical was identified, the industry that was most likely to be discharging it was contacted on an informal basis ... the waste stream was immediately taken out of the river and the chemical has not reappeared" [3]. A decade later, this description of ONCB in the Mississippi River helped to catalyze a new environmental law. In its 1971 report *Toxic Substances*, the President's Council on Environmental Quality cited the data from this study to illustrate the need for Congress to pass the Toxic Substances Control Act (TSCA) [7]. And decades after that, the industrial site which apparently discharged ONCB became a Superfund site [8]. The US EPA has also regulated discharges of chlorinated benzenes to surface water under the Clean Water Act (40 CFR 401.15) and emissions to air from the manufacture of ONCB (40 CFR 401.15) under the Clean Air Act.

Market forces are also at work. As of 2015, ONCB was no longer manufactured in the US. A single company imported the substance, in a quantity of less than 25,000 pounds (approximately 11,300 kilograms) per year [9].

The fate of any residual ONCB in the environment from historical releases can be predicted from the physical–chemical properties of the substance. ONCB would not likely persist in water or soil, but has the potential to persist in sediment and to be transported long distances by air. It would not be expected to bioaccumulate [10].

This example illustrates several important points relevant to the notion that we find what we look for:

- The scientific and regulatory tools that allow us to identify emerging contaminants and put them in context are relatively young. As they evolve, society has identified "new" contaminants. The environmental contaminants we now investigate and remediate reflect our ability to identify a chemical in the environment, recognize the resulting hazards, and determine the need to regulate. Regulators have largely built lists of target chemicals from the "bottom up" as concerns about specific chemicals have arisen.

- The evolving science of toxicology and availability of new test data change our understanding of hazard and risk, and as a result the level of exposure to a chemical that regulators are willing to accept. Early on, contaminants were often designated as harmful based on acute hazards or the potential to cause cancer. Over time and as we'll return to in a later chapter, scientists have begun to identify and measure more subtle effects such as potential endocrine disruption, neurotoxicity, or reproductive effects. Such findings can cause a chemical to emerge as a new contaminant of concern.
- Emerging issues can be fleeting. As an emerging contaminant, outrage over ONCB in the environment led to a new environmental law on the manufacture and use of chemicals in the US. But industrial trends have changed, decreasing the manufacture and use of ONCB, and environmental concentrations have attenuated over time. As the example of ONCB illustrates, contaminants have often emerged as our society has become aware of hazards or risks of a chemical substance. We'll return to the implications of such awareness for "new" contaminants to emerge in Chapter 3. But first, we'll examine the factors that determine how we have identified contaminants in the past and how those factors will likely influence actions in future.

2.2 EVOLUTION OF ENVIRONMENTAL REGULATIONS AND DEFINITION OF CONTAMINANTS

Society's definition of contamination has evolved over time. The earliest environmental legislation in the US, the Rivers and Harbors Appropriation Act of 1899, defined unacceptable contamination of navigable waters as "any refuse matter of any kind or description whatever other than that flowing from streets and sewers" [11]. Even as late as the mid-1900s some accepted water pollution as a natural consequence of industrial development. From that perspective, some saw events such as the Cuyahoga River fires of the 1950s and 1960s as an acceptable risk outweighed by the economic benefit of heavy industry [12].

When the US EPA originated in the early 1970s, it focused on gross pollution problems. The Agency often lacked the data needed to assess a specific contaminant or fully understand the regulatory consequences [13]. In the early days, regulations also focused on human health impacts rather than potential ecological impacts [14]. Major environmental legislation passed in the 1970s and 1980s, and the pollutants designated under those laws, formed the basis for identifying contaminants at uncontrolled hazardous waste sites. Landmark legislation in the United States included the following.

- Clean Water Act of 1972 (CWA; based on the FWPCA of 1948) [15]: as described at 33 U.S.C. 1251 Sec. 101, "the objective of this Act is to restore and maintain the chemical, physical, and biological integrity of the Nation's waters." Among other provisions, the law prohibits the discharge of pollutants into navigable waterways in toxic amounts; this stricture can pertain to permitted industrial wastewater discharges and to spills or leaks of hazardous substances. Under the CWA, the US EPA initially created two related lists of chemicals, the Toxic Pollutant List and the Priority Pollutant List. The Toxic Pollutant List, codified at 40 CFR 401.15, comprises 65 substances or groups of substances. Priority Pollutants are the 129 chemicals that US EPA regulates as pollutants under the CWA. The Priority Pollutant List includes chemicals that meet four criteria: substances on the Toxic Pollutant List; an analytical method existed at the time the list was developed; the chemical had been found in samples tested with a frequency of occurrence of at least 2.5%; and the chemical was produced in significant quantities according to the Stanford Research Institute's *1976 Directory of Chemical Producers, USA*. The US EPA has not updated these lists since 1977, although the Agency can develop effluent guidelines for pollutants not on these lists [16]. The US EPA also maintains a Hazardous Substances List under the CWA. Regulations in 40 CFR 117.3 list substances designated as hazardous under Section 311(b)(4) of the CWA. Releases of a hazardous substance over its Reportable Quantity triggers reporting requirements and liability for removal. The list, last amended in 1995, comprises 296 substances [17].

- Clean Air Act [18]: This law, passed in 1970 and amended in 1977 and 1990, requires the US EPA to establish national ambient air quality standards for certain common and widespread air pollutants. These standards form the basis for enforceable plans developed by each State. The law also requires US EPA to regulate toxic air pollutants, i.e., pollutants that are known or suspected to cause cancer or other serious health effects, such as reproductive effects or birth defects, or adverse environmental effects. As of the 1990 Amendments, this list includes 187 hazardous air pollutants [19].
- Resource Conservation and Recovery Act (RCRA) [20]: This 1976 law, amended in 1984, governs the management of solid waste and, more specifically, the generation, transportation, treatment, storage, and disposal of hazardous waste. Regulations at 40 CFR 261 Part D list specific hazardous wastes.
- Toxic Substances Control Act (TSCA) [21]: Originally passed in 1976 and amended in 2016, this law regulates the manufacture and use of chemicals in the United States. It gives the US EPA the authority to restrict the manufacture, use, or disposal of chemicals which present an unreasonable risk. As discussed later in this book, the TSCA Inventory currently contains over 40,000 chemicals in commerce in the US.
- Safe Drinking Water Act (SDWA) [22]: This 1974 law protects the quality of public water supplies. Under the SDWA, the US EPA sets National Primary Drinking Water Regulations (NPDWRs) and

National Secondary Drinking Water Regulations (NSDWRs). NPDWRs pertain to specific microorganisms, disinfectants, disinfection byproducts, organic and inorganic chemicals, and radionuclides. The organic chemicals include 53 substances or groups of related substances; the list of regulated inorganic chemicals comprises 16 substances [23]. NSDWRs are unenforceable guidelines that the US EPA has set for 15 contaminants that can present an aesthetic nuisance, such as taste, color, and odor [24]. The Unregulated Contaminant Monitoring Rule (UCMR) promulgated under the SDWA Amendments of 1996 gives the US EPA the authority to collect data for contaminants that do not have health-based standards set under the SDWA and that the Agency suspects to be present in drinking water. The Contaminant Candidate List (CCL), as periodically updated, designates the chemicals to be monitored. We'll return to the UCMR and CCL in Chapter 3.

These regulatory programs influenced the development of legislation and regulations on uncontrolled hazardous waste sites. The Comprehensive Environmental Response, Compensation, and Liability Act of 1980, abbreviated CERCLA and better known as "Superfund", ignited the practice of hazardous waste site investigation and cleanup and served as the model for related State laws in the United States. The US House of Representatives described the intent of the bill in the title of H.R. 7020, a precursor to the final legislation: "An act to provide for liability, compensation, cleanup, and emergency response for hazardous substances released into the environment and the cleanup of inactive hazardous waste disposal sites".

But what did Congress consider to be a hazardous substance? The final bill [25] defined "hazardous substance" largely based on chemicals identified in preceding legislation: any hazardous substance or toxic pollutant designated by the FWPCA (now known as the CWA); any hazardous waste identified under or listed pursuant to specified provisions of the Solid Waste Disposal Act (better known now as RCRA); any hazardous air pollutant listed under specified provisions of the Clean Air Act; and any imminently hazardous chemical substance or mixture as defined under TSCA. The bill also allowed the US EPA to designate a substance or mixture as a hazardous substance under Section 102(a) of CERCLA if such substances "when released into the environment may present substantial danger to the public health or welfare or the environment." Regulations at 40 CFR 302.4 list the hazardous substances designated under CERCLA. These hazardous substances, derived largely from lists of chemicals developed for regulation under other environmental laws, have formed the basis for site investigation and remediation.

As the CERCLA program proceeded, observers soon realized that the list of hazardous substances was far from complete relative to the chemicals in commerce that could be of environmental concern. Writing in 1984 about the National Contingency Plan (the regulations which implement CERCLA), one analyst commented [26]:

Existing regulations frequently will not apply to significant portions of the hazards that the agency or party seeks to abate ... most of these regulations specifically address the health effects of relatively few substances ... there are thus two sources of uncertainty in using other environmental acts to formulate standards for the necessity and adequacy of responses under the National Contingency Plan: the uncertainty concerning the health effects of many of the substances regulated, and the factual uncertainty in predicting future exposures.

Within a few years a catastrophic chemical release proved this writer's assessment to be true. A train car derailment and release of the chemical metam sodium in 1991 led the US EPA to conclude that the list of hazardous substances under CERCLA was inadequate and to convene a Hazardous Substances Task Force. The Task Force reported that [27] (emphasis added):

On July 14, 1991, several Southern Pacific railway cars derailed on the Cantara Loop near Dunsmuir, California. One of those railway cars ruptured and released approximately 19,500 gallons of the herbicide metam sodium into the Sacramento River. As a result of the spill, the surrounding environment along a 45-mile stretch of the river and portions of Lake Shasta were significantly affected, more than 200,000 fish were killed, and several hundred people were treated for eye, skin, and respiratory irritation. Metam Sodium is not a listed hazardous substance under the Comprehensive Environmental Response, Compensation, and Liability Act (CERCLA), nor was the metam sodium shipment regulated by the U.S. Department of Transportation ... However, its release was subject to CERCLA notification and liability provisions because the metam sodium rapidly hydrolyzed and decomposed into at least three specifically listed CERCLA hazardous substances, and because the released substance became a hazardous waste under the Resource Conservation and Recovery Act (RCRA) ... [this situation] points to the fact that many chemicals that are potentially hazardous to human health and the environment are not specifically listed under CERCLA ... even though such chemicals may pose a significant hazard upon release. *As such, the spill raised public concerns that the regulatory framework of the Federal government often overlooks many chemicals that are, or potentially could be, hazardous to human health and the environment.*

Among other recommendations, the Task Force advised further work to develop screening criteria for hazardous substance designation or take other actions to refine or expand upon the work.

A search of US EPA's web site and more generally of the internet in February 2019 did not turn up evidence that the US EPA acted upon the recommendations of the Hazardous Substances Task Force in a way that broadly influenced the designation of hazardous substances under CERCLA. After a report to Congress in 1992 [28,29] and a scoping study of hazardous waste characteristics under RCRA [30] that cited the Task Force report, published reports of the Task Force and their work appear to cease. As a result, the investigation and remediation of hazardous waste sites largely focus on lists of target compounds that have not been updated in decades.

In summary, regulatory lists of hazardous substances are a lagging indicator of concern. Almost by definition, regulations reflect known hazards from substances that have already been detected in the environment. The lists of regulated chemicals, many of which originated decades ago, change slowly, if at all. Until society becomes aware of exposure and resultant risks, we are unlikely to regulate contaminants and thus the potential exists for an unlisted contaminant to emerge. In the next two sections, we will examine two factors relevant to the identification of a "new" contaminant: the availability of analytical techniques and our understanding of hazard.

2.3   DRIVEN BY ANALYTES

The chemicals we can see in environmental media depend entirely on our analytical methods and what they are designed to detect along with their associated method detection limits. When we anticipate developments in emerging contaminants, it is important for us to recognize that we only see what we are looking for.

One of the complicating factors in analytical chemistry is that the chemicals we can detect and the levels at which we can detect them in various media will vary across sample preparation methods and analytical techniques. This is why regulators have specified standardized analytical methods for different types of compounds. These standardized methods limit our view of contamination: they generally reflect target lists of contaminants developed from existing regulations. As described in the preceding section of the book, the listing of chemicals in regulations necessarily lags behind the evolution of environmental concerns.

We can learn about how we define what we are looking for, with respect to contamination of soil and water, by examining very briefly the evolution of environmental analytical chemistry.

For typical soil and groundwater organic contamination, analysis starts with extraction of chemical compounds from the sample itself, based on either the chemical's volatility or solubility in a solvent. The next step is often chromatography, a method that separates the components of the chemical mixture extracted from a sample based on the physical/chemical properties of each component chemical.

Chromatography enables a detector to "see" and quantify individual compounds. The science of chromatography started back in the early 1900s when Mikhail Tswett, a Russian scientist, separated plant pigments in a liquid column [31]. Despite that early beginning, the methods used in environmental analytical chemistry are still relatively young. Several different types of chromatography evolved from this discovery with the advent of gas chromatography (GC) in the early 1950s [32]. Most of today's standard analytical methods used for organic compounds in soil and groundwater include the use of a GC coupled with mass spectrometry (GC/MS) to identify chemical compounds.

GC/MS analyses generally rely on software associated with commercial analytical equipment that contains a library of reference compounds for each specified test method. If a chemical compound is not in the reference library for the specified test method, the chemical may not be identified in a sample unless the laboratory compares the MS output to a larger universe of reference compounds. The use of GC/MS enables a laboratory to identify up to 250,000[2] compounds [33], but the full power of the tool is not commonly used. Consider, for example, that routine analyses of organic contaminants in the US often rely on two workhorse methods. US EPA Method 8260 (or some modification of this method), used to evaluate volatile organic compounds in soil and groundwater, can detect approximately 100 different compounds [34]; US EPA Method 8270 (or modification) for semi-volatile organic compounds focuses on approximately 250 substances [35]. Thus, routine analyses using these methods may detect only a fraction of the organic chemicals present in a sample unless the laboratory is instructed to utilize the full library of 250,000 reference compounds.

The standard general approach to identifying contaminants provides a needed and reliable source of data; however, it does not characterize all the chemical components of a sample. Consider petroleum, a complex mixture of over 100,000 components [32]. (This example is included to illustrate the limitations of common analytical methods, not to explicitly suggest that petroleum contains an emerging contaminant.) Gasoline and diesel oil each comprise four main structure classes of hydrocarbons: n-alkanes (linear saturated hydrocarbons), isoalkanes (branched saturated hydrocarbons), cycloalkanes (saturated cyclic alkanes), and aromatics [36]. Gasoline comprises a mixture of these 4 classes that contain 4 to 10 carbons, while diesel contains hydrocarbons with 11 to 25 carbons [36]. Gasoline contains approximately 230 different hydrocarbons which can be isolated by a GC [36], while diesel contains a much higher number of hydrocarbons – estimated to be between 2,000 and 4,000 – and they cannot be fully separated by GC.

What are we looking for in our standard analytical lists used at petroleum sites? We typically use a number of methods. For gasoline, it is common to run US EPA Method 8260 (or some modification of this method) to evaluate volatile organic compounds in soil and groundwater. As described above, this method can detect approximately 100 different compounds, only a portion of volatile organic gasoline components. Some of these compounds can be found in gasoline, but many are not related to petroleum substances (i.e., chlorinated compounds, ketones, alcohols). For diesel, it is common to run US EPA Method 8270. This method can detect approximately 250 compounds, yet it identifies nowhere near the number of compounds thought to be present in diesel (2,000 to 4,000 [36]).

To overcome some of these gaps in standard analytical methods and to provide a cost effective way to evaluate the extent of gasoline or diesel in environmental media, chemists have developed methods such as Total Petroleum Hydrocarbons – Gasoline (TPH-G) or TPH – Diesel (TPH-D). However, these are qualitative tests that provide only an aggregate indictor of all the TPH-G or TPH-D present, and do not provide any indication of the individual compounds present or their concentration in the environmental media. This leads to a host of questions: What are these compounds? Should we be concerned about them? Do they cause a risk? In fact, are we sure that what we are seeing in the results is actually petroleum? This is an example of how petroleum, while studied in the environment in the US for decades, is still subject to analytical issues in terms of quantification and evaluating risk. Moreover, it illustrates how the use of routine analytical methods may miss some (or many) chemical compounds. We find what we look for and as a result, the use of standard methods to characterize environmental conditions creates the potential for "new" contaminants not previously seen in the data to emerge from existing conditions.

All analytical methods also have a practical quantification limit, meaning there is a lower bound concentration to what can be measured in a sample using a specific method. All chromatography methods require running calibration curves to quantify compounds and relate the output of the detector to a specific concentration. The lower and upper bounds of the calibration curve frame the detector response for which there is confidence in relating response to concentration. The practical quantification limit is the lowest detector response where there is still a high level of confidence in concentration relationship. Each method also has a minimum detection limit, which is the smallest detector response that the method can reliability detect. It is therefore incorrect to assume that because the analysis does not detect specific constituents, it means they are not present (they may be, but at a lower level than can be detected), or that nothing else is in the sample beyond the list of identified analytes.

This last point, that a sample may contain chemical compounds not detected because they are not among the analytes recognized by standard methods, is particularly important with respect to emerging contaminants. Many readers will have seen laboratory reports indicating unknown or so-called "tentatively identified compounds" (TICs) in an environmental sample. A laboratory may identify TICs not found in routine analyses if it is instructed to consult the larger reference library of some 250,000 compounds. It may also identify TICs if, based on specific instructions, chemists have manually assessed the analytical output to identify chemicals or classes of chemicals rather than analyze only the compounds specified in the analytical method.

As a practical matter, TIC data have often been unusable. In order to gauge the risk of exposure to a chemical and thereby focus investigation and remediation efforts, one must understand the hazards presented by exposure to a TIC. Historically such hazard data have not been available. As described in Section 1.2, the US EPA's Integrated Risk Information System (IRIS) database contains scientific positions on human health effects from only 568 contaminants. This scenario is changing, however, as revisions to global chemical regulations have spurred the development of toxicity data for many of the chemical compounds we could not assess in the past. We'll explore this important point in Sections 2.4 and 3.2 of this book.

In summary, while we have the potential to quantify chemicals in environmental media, the chemicals we see and the concentrations at which we can see them entirely depends on the analytical method. Standard methods allow for routine and reliable analysis, but may miss contaminants omitted from regulatory lists. Thus, we are limited in our ability to fully understand environmental contamination simply through the analytical methods we typically employ at hazardous waste sites.

It is often not until after a contaminant has emerged as a potential issue that chemists develop analytical methods. Consider PFAS, a class of compounds with more than 3,000 fluorinated organic chemicals [37]. At the time of this writing in 2019, the US EPA has developed one analytical method for analyzing PFAS in drinking water (US EPA Method 537.1), which can detect 18 PFAS compounds, with an additional drinking water method capable of detecting 25 PFAS compounds under development. The US EPA estimates a non-drinking water method will be finalized for PFAS in the fall of 2019 [38]. In addition, several methods are under development by various laboratories, using GC/MS or liquid chromatography-mass spectrometry-mass spectrometry (LC/MS/MS) [39]. Laboratories have the ability to quantify only a handful (20+) of the PFAS compounds to the low levels being use for evaluation, i.e., parts per trillion. This raises the question of whether there are additional PFAS compounds present in media, or other compounds not currently under scrutiny, at these extremely low concentrations. When contemplating the part per trillion level, there are likely many more compounds present in environmental media that we currently aren't looking for, we don't have risk data on, and we don't understand enough to know whether they are of concern. But the picture is changing; in the sections below, we will explore the evolution of our ability to understand the hazards presented by environmental contaminants.

2.4   UNDERSTANDING HAZARDS AS A BASIS FOR ACTION

Risk to human health or the environment is a function of exposure and hazard. A significant risk to human health or the environment prompts action to characterize and remediate contamination. As our understanding of the hazards associated with a certain chemical or groups of chemicals continues to evolve, new contaminants can emerge.

Data that characterize the hazards of chemical exposure have historically come from several sources: studies performed by academics; testing conducted by governmental entities such as the US Department of Health and Human Services under the National Toxicology Program; or from industry seeking to bring new chemicals to market under TSCA or analogous requirements. Data from the first two sources, logically enough, are often generated when scientists believe a hazard may exist and can consequently secure the funding to perform testing. Thus, a contaminant must have emerged in some respect before such tests are performed. The third source of data, from testing required to bring chemicals to market, reflects two factors. First, tens of thousands of chemicals were in commerce before legislators passed chemical control laws that required testing of toxicity or ecotoxicity. Until relatively recently, most of those chemicals were not tested and therefore data regarding toxicity were limited. Second, even testing of chemicals brought to market since legislation passed in the 1970s was relatively limited until circa 2008. The subsections below examine each of these two factors.

2.4.1 Data Generated to Bring Chemicals to Market[3]

The regulation of the manufacture and use of chemicals under TSCA in the US provides important perspective on the limitations on understanding the toxicity and ecotoxicity of hazardous substances in the environment. Until 1976, when Congress passed TSCA, the United States did not regulate the manufacture and use of most chemicals. By that time, and as illustrated in Figure 2.1[4] the chemical industry had been established for over a century. Approximately 62,000 chemicals were reported to be in commerce [40], and chemical production in the United States was worth nearly $455 billion (adjusted to 2010 dollars) [41,42].

TSCA designated chemical substances manufactured in or imported into the US before January 1, 1975 as existing chemicals. The original TSCA Inventory listed approximately 62,000 existing chemicals. Over the next 40 years, the TSCA Inventory grew to list more than 84,000 chemical substances, based on the submittal of Premanufacture Notices (PMNs) of the manufacture (or importation) of new chemicals. (The US EPA determined in 2019 that 40,655 chemicals were then in commerce and should be considered "active" on the TSCA Inventory [43].)

The US EPA made some effort to assess the hazards and risks of existing chemicals on the original TSCA Inventory. In an initiative begun in 1998, the Agency focused on High Production Volume (HPV) chemicals, i.e., those manufactured or imported in more than 1 million pounds (approximately 450,000 kilograms) per year. The program included approximately 2,800 existing chemicals [44]. At that time no data were publicly available on the basic toxicity of 43% of HPV chemicals. Further, only 7% of HPV chemicals had

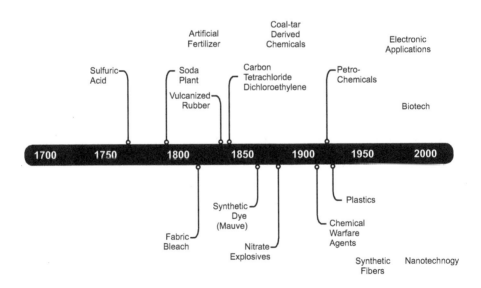

FIGURE 2.1   Selected Milestones and Trends in the Modern Chemical Industry

data available for all six of the endpoints that US EPA considered to be fundamental: acute toxicity, repeat dose toxicity, developmental and reproductive toxicity, mutagenicity, ecotoxicity, and environmental fate [45].

In this voluntary HPV program, manufacturers and importers of HPV chemicals sponsored the development of screening-level health and environmental data. As of July 2013, the US EPA had amassed over 340 submissions representing almost 900 chemical substances, either as a single chemical submission or as a member of a chemical category. The Agency evaluated the toxicity data for 229 of these substances/categories and published a report on each to prioritize HPV chemicals for follow-up data collection or management actions based upon their potential risks [46]. While these data and assessments provided valuable information, they addressed only a fraction of the existing chemicals in use in the United States.

Of the approximately 22,000 new chemicals added to the TSCA Inventory based on PMNs between the late 1970s and 2016, only a fraction were brought to market with extensive test data on the potential hazards to human health or the environment. TSCA required that PMNs must only include "all existing health and environmental data in the possession of the submitter, parent company, or affiliates, and a description of any existing data known to or reasonably ascertainable by the submitter" (40 CFR 720.40(d)). The US EPA could not require a manufacturer to collect any additional data on the hazards of a chemical substance unless it could present an "unreasonable risk". The US EPA found in a survey of the PMNs submitted between 1979 and 1985 that 44% of the 5,500 submittals contained some toxicological data (typically on acute toxicity) and 9% contained some ecotoxicological data (also typically on acute toxicity). Fifty-four percent contained no test data. Trends circa 1997 were similar [47]. Although the US EPA later requested data on some of these chemicals, the overall trend was that few data were available to quantify the hazards of chemical substances in commerce in the United States.

Such limitations on data available to characterize the hazards of chemicals in commerce were not confined to the United States. The European Union, for example, recognized that the chemical control law in place before 2008 "often proved itself to be incapable of identifying risks posed by many chemicals and was slow to act when risks were identified" [48].

Changes in the chemical control laws in the United States and European Union between 2008 and 2016 have resulted in an extraordinary effort to assess the hazards and risks of the chemicals in commerce. The resulting data will revolutionize our understanding of environmental contamination and cause new contaminants to emerge into awareness. Chapter 3 provides an analysis of these potential effects.

2.4.2 TOXICOLOGY AND ECOTOXICOLOGY: EVOLUTION OF TEST METHODS

The practice of toxicology and ecotoxicology within a regulatory setting is still relatively young, as illustrated by the case study of ONCB in Section 2.1. One measure of the ability to identify adverse effects on human health or environmental receptors is the availability of standardized test methods to evaluate the toxicity of a chemical. Following is a brief review of the development of test methods in the United States and globally. As test methods are published and then come into common use, the understanding of hazard grows and the potential develops for concerns about contamination to emerge.

2.4.2.1 Efforts in the United States

The evolution of US EPA's guidelines on hazard and risk assessment shows the Agency's evolving focus on the effects of exposure to environmental pollutants. US EPA's first foray into formally outlining an approach for evaluating the effects of chemical exposure on human health was the publication of *Interim Procedures and Guidelines for Health Risk and Economic Impact Assessments of Suspected Carcinogens* in 1976 [49]. This document provided a brief summary of US EPA's overall approach for addressing chemicals that were suspected of causing cancer in humans and was primarily focused on identifying and regulating carcinogenic pesticides. US EPA identified the use of epidemiological studies with support from animal testing as the best methods to determine if a chemical could cause cancer with elevated levels of exposure. In order to estimate the carcinogenic potential of a chemical, US EPA focused heavily on rodent bioassay tests that identified cancer endpoints in test animals. Although the 1976 Guidelines speak to considering a weight of evidence approach, US EPA appeared to strongly focus on assuming a non-threshold dose–response relationship between exposure and effects to ensure that extrapolations between animal studies and humans were sufficiently protective [50].

Over the next 10 years, US EPA continued to focus on cancer endpoints and refine its approach to carcinogenic risk assessment. In 1986, US EPA published the *Final Guidelines for Carcinogen Risk Assessment* [51]. These updated guidelines were partially based on *Chemical Carcinogens: A Review of the Science and its Associated Principles*, February 1985 [52], and provided a framework for establishing a series of general principles for assessing carcinogenic risk. Indeed the overall goal of the 1986 *Guidelines* was to elucidate the basic mechanisms underlying cancer causation and provide an approach for making sound and reasonable regulatory judgments. Topics covered in the 1986 *Guidelines* included the mechanisms of carcinogenesis, short-term testing and the relationship between genetic toxicity and carcinogenicity, evaluation of long-term animal tests, and the current state of epidemiological data as related to cancer in humans.

It was not until 1986 that US EPA would publish its first guidance on evaluating noncancer health effects in humans. The *Guidelines for the Health Assessment of Suspect Developmental Toxicants* [53] provided guidance on assessing and evaluating developmental toxicity in humans where cancer is not the primary mechanism. The 1986 *Guidelines* were quickly updated in 1991 with the *Final Guidelines for Developmental Toxicity Risk Assessment* [54]. The 1991 *Guidelines* expanded US EPA's guidance on noncancer health effects related to developmental toxicity. Despite the focus on carcinogenicity, by 1991 the IRIS database contained 371 chemicals, of which 43 had a cancer toxicity value and 268 had a noncancer toxicity value. (Note: This is based on data available from the IRIS database in 2019 and may not reflect historical data if a chemical was recently updated).

Another 10 years would pass before US EPA would again propose to update its guidelines for evaluating cancer in humans. The *Proposed Guidelines for Carcinogen Risk Assessment* were released in 1999 and underwent several rounds of internal and external review before they were updated in 2005 [55,56]. At the same time, US EPA began to work on guidelines for evaluating noncancer health effects and in 2002 published *A Review of the Reference Dose and Reference Concentration Processes* [57]. For both these documents, US EPA focused on incorporating data and approaches to account for addressing potential health effects to children. US EPA's focus on children's health began in 1995 with the release of the *New Policy on Evaluating Health Risks to Children* [58]. In this policy, US EPA explicitly identified risks to infants and children as a critical part of any risk assessment and indicated that as appropriate, separate assessments would be conducted to evaluate risks to infants and children. Most recently, in 2013, US EPA reaffirmed its policy on, and committed to continuing to protect, children's health [59].

While human health was initially the focus for US EPA, ecological effects were also an area of concern to state and Federal regulators. The earliest efforts to characterize hazards to aquatic life included compilations of the scientific literature on aquatic toxicology and corresponding water quality criteria by the State of California in 1952 and 1963; National Technical Advisory Committee to the Secretary of the Interior in 1968; and the National Academy of Sciences (on behalf of US EPA) in 1972. The Agency published the "Red Book" of water quality criteria under the CWA in 1976 [60], and the follow up "Gold Book" in 1986. The criteria include values intended to protect aquatic life.

US EPA first published guidance on evaluating risks from ecological receptors in 1992 [61]. The *Framework for Ecological Risk Assessment* relied on information discussed and presented in the *Summary Report on Issues in Ecological Risk Assessment* [62]. Overall, the goal of the *Framework* was to foster consistent approaches to ecological risk assessment through the use of a basic structure and starting principles. US EPA updated the *Guidelines for Ecological Risk Assessment* in 1998 to provide additional detail and structure to the process for conducting ecological risk assessments [63].

US EPA continues to refine and update guidance documents related to the evaluation of human and ecological health. This includes new guidance to address receptors for pesticides (e.g., workers and residents) [64,65], endpoints (e.g., neurotoxicity and immunotoxicity) [66], and methods (acute toxicity and benchmark dose guidance) [67,68] to name just a few. These guidance documents provide more detailed direction and guidance for US EPA to evaluate chemical toxicity and as a result may lead to the identification of chemicals with health effects that had not been previously known.

Global efforts at characterizing the hazards of chemical exposure, to which US EPA has contributed, have also grown during the last few decades.

2.4.2.2 Global Efforts

On a global level, the Organization for Economic Cooperation and Development (OECD) member and partner countries have developed the OECD Guidelines for the Testing of Chemicals [69]. These guidelines include five series of tests:

- 100 series: Physical Chemical Properties
- 200 series: Effects on Biotic Systems (i.e., daphnia, fish, birds, earthworms, bees, and other species)
- 300 series: Environmental Fate and Behaviour
- 400 series: Health Effects (i.e., tests intended to characterize the effects of chemical exposure on human health)
- 500 series: Other Test Guidelines.

Figure 2.2 illustrates the development of new guidelines since the program began in 1981. Efforts began with just 20 guidelines for measuring the hazards of chemical exposure. Those early guidelines included tests on freshwater algae and cyanobacteria (growth inhibition), *Daphnia sp.* (acute immobilization), and fish (acute toxicity). They also included five tests designed to evaluate the potential for acute toxicity in humans as well as guidelines on measuring repeated dose/subchronic effects, and a measure of carcinogenicity. Starting in 1983, OECD expanded the categories of data to be evaluated and released test methods on genetic toxicology testing.

As illustrated in Figure 2.2, OECD has continued to publish updated test methods to cover a fuller range of biotic system tests (e.g., vegetative, fish, and honeybee tests), and health effect methods (e.g., estrogen receptors). OECD has

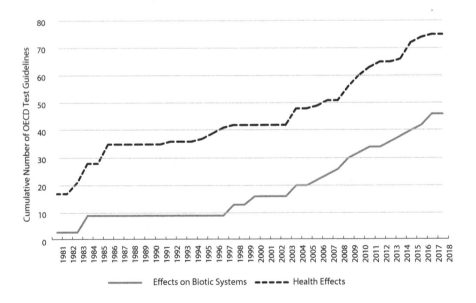

FIGURE 2.2   Development of OECD Test Guidelines

also expanded the suite of guidelines to include more species in the environment, and to provide more tools to assess various effects on human health. These updated and expanded test methods reflect the expanded knowledge about chemical toxicity.

For example, as noted above, OECD added test methods for estrogen receptors in 2007, 2009 and 2012. These test methods reflected an increase in our understanding of endocrine effects. The *State of the Science of Endocrine Disrupting Chemicals – 2012* [70] detailed additional research and data collected on endocrine disrupting chemicals, including the finding that effects in humans are very similar to those seen in wildlife, especially if humans are exposed at a sensitive or vulnerable life stage. Indeed, these findings contradict earlier reports which suggested that the effects seen in wildlife were not similarly occurring in humans and that there was only weak evidence for endocrine effects in humans [71]. The potential for endocrine disruption is just one of the hazards and corresponding sets of test methods that have evolved since the 1980s, with the potential for a different understanding of hazard that can cause a "new" contaminant to emerge as data become available. We'll explore how regulatory developments drive the collection and assessment of new toxicity data in Chapter 3.

In summary, the understanding of the toxicity of many chemicals has been limited by the lack of availability of data, due in part to the historic absence of regulatory requirements to collect data. As discussed in this chapter, consistent test methods were not historically available to evaluate chemicals used in industry or the methods and approaches to evaluate test data were not readily available. Not until the mid-1970s was there a consistent approach to collect and evaluate toxicity data and make determinations about chemical exposures in humans and eventually ecological receptors.

2.4.3 EVOLUTION OF IRIS

Regulators have compiled dose–response data for a relatively small number of environmental contaminants. While such regulatory databases have supported the cleanup of specific pollutants at thousands of sites, they omit many other potential contaminants. To provide context for the emergence of "new" contaminants, we'll examine the strengths and limitations of one of the foundational databases.

The US EPA formalized its views on the dose–response data that would drive site investigation and remediation in the mid-1980s, when it created IRIS [72]. This database, maintained as an internal resource at that time, contained dose–response factors representing Agency consensus scientific positions on human health effects that could result from chronic oral, dermal, or inhalation exposure to chemicals in the environment. In 1987, US EPA scientists began to add summaries of relevant data to the database; in 1988, the Agency made IRIS publicly available. IRIS continues to evolve and grow, with the addition of new supporting documentation and the inclusion of new chemical substances.

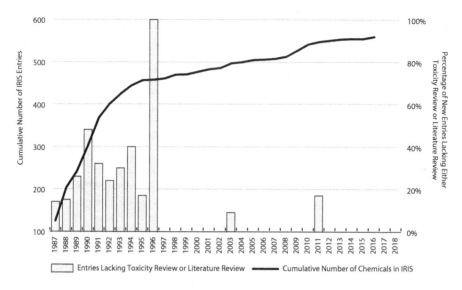

FIGURE 2.3   Evolution of IRIS

As illustrated in Figure 2.3 [73], most IRIS entries originated in the 1980s. Those entries often focused on limited toxicological effects and lacked supporting scientific information. Of the 568 entries into IRIS, 124 appear to lack supporting data in the IRIS database [73].

Consider, for example, 1,2-dichloroethane (abbreviated 1,2-DCA; CAS 107-06-2). US EPA created an IRIS entry for this chemical in 1987 [74]. The US EPA determined that 1,2-DCA was a probable human carcinogen based on a 1978 study. The Agency did not evaluate the risks from noncancer effects. Three decades later, this work still guides site investigation and remediation including driving vapor intrusion investigations despite information in the Registration, Evaluation, Authorisation and Restriction of Chemicals (REACH) database which indicates that the information in the IRIS database is outdated [75].

Even more recent entries into IRIS have raised concerns regarding their scientific rigor. In 2011, the National Academy of Sciences wrote this about the US EPA's decade-long review of formaldehyde toxicity data and the draft IRIS assessment report [76]:

> Problems with clarity and transparency of the methods appear to be a repeating theme over the years, even though the [assessment reports] appear to have grown considerably in length. In the roughly 1,000-page draft reviewed by the present committee, little beyond a brief introductory chapter could be found on the methods for conducting the assessment. Numerous EPA guidelines are cited, but their role in the preparation of the assessment is not clear. In general, the committee found that the draft was not prepared in a consistent fashion; it lacks clear links to an underlying

conceptual framework; and it does not contain sufficient documentation on methods and criteria for identifying evidence from epidemiologic and experimental studies, for critically evaluating individual studies, for assessing the weight of evidence, and for selecting studies for derivation of the [reference concentrations] and unit risk estimates.

The National Academy of Sciences concluded that between 2011 and 2018 the US EPA improved the quality of IRIS data [77]. However, limitations remain on the number of chemicals represented in the database and the nature and quality of the early entries.

2.5   IDENTIFYING CONTAMINANTS AND THEIR HAZARDS

Our ability to understand hazards as a basis for identifying contaminants of concern evolved slowly. Consider this chronology of regulators' efforts to quantify hazards:

- Circa 1970: The Occupational Safety and Health Administration (OSHA) undertook one of the earliest efforts to quantify hazard in a regulatory context when it derived Permissible Exposure Limits (PELs) for workers who were manufacturing or using chemicals circa 1970. PELs largely reflected the 1968 Threshold Limit Values developed by the American Conference of Governmental Industrial Hygienists. Since this initial effort, PELs have not been updated. OSHA acknowledged in 2019 that "[i]ndustrial experience, new developments in technology, and scientific data clearly indicate that in many instances these adopted limits are not sufficiently protective of worker health" [78].
- Circa 1976: The US EPA published the "Red Book" of water quality criteria under the CWA in 1976, and the follow up "Gold Book" in 1986. The criteria include values intended to protect aquatic life and values developed to protect human health. Some of these criteria are still in use [79]. They can form the basis for cleanup goals at contaminated sites.
- 1985: US EPA created the IRIS database, which as of June 2019 contained dose–response information for 568 chemicals. As described earlier in this chapter, many of these assessments reflect decades-old data which are poorly documented in IRIS.
- 2006–present: Under the European Union regulation REACH, chemical manufacturers and importers in the European Union had to compile toxicological and ecotoxicological data on chemicals in commerce. This requirement generated unprecedented amounts of information about the hazards of exposure to chemicals. As of June 2019, the European Chemical Agency's database of registered substances contained data on approximately 25,000 chemical compounds [80]. This cache of data dwarfs the information compiled by US EPA under CERCLA, the CWA, TSCA, and other environmental laws. The REACH database includes new data on chemicals that US EPA has evaluated in the past,

and data on chemical compounds which are in commerce, but have not been assessed for toxicity by US EPA.

- 2019: As noted above, US EPA's efforts to catalog chemicals in commerce have determined that 40,655 chemicals are made in or imported into the United States. Under the 2016 amendments to TSCA, the Agency must assess the potential hazards and risks from all high-priority chemicals.

We'll explore the final two points in this chronology in Chapter 3. They're included here to provide perspective on this concept. We find what we look for. Site investigation and remediation often focus on lists of chemical compounds identified decades ago and assessed, sometimes, based on limited data on their hazards. Efforts under REACH and TSCA within the last decade provide new perspective on hazards and risks that may upend that old thinking. As a result, we're now at a pivotal point, where new data are becoming available and so the old lists of contaminants can increase.

NOTES

1. This case study is adapted from Section 4.1 of *Product Stewardship, Life Cycle Analysis and The Environment* (Taylor & Francis/CRC Press, 2015) and is used with permission.
2. Readers may be confused by the ability to analyze 250,000 organic compounds when this book refers to approximately 40,000 chemicals in commerce in the United States. Many chemical compounds are used in research and development or otherwise in small quantities, and therefore are not counted as chemicals in commerce.
3. This section is adapted from Section 3.1.2 of *Product Stewardship, Life Cycle Analysis and the Environment* (Taylor & Francis/CRC Press, 2015) and is used with permission.
4. This figure is adapted from Figure 1.1 of *Product Stewardship, Life Cycle Analysis and the Environment* (Taylor & Francis/CRC Press, 2015) and is used with permission.

REFERENCES

1. Green, R.S., 1963. Data gathering and monitoring equipment in water supply and water pollution control programs. *American Journal of Public Health and the Nation's Health*, 53(12), pp. 1963–1971.
2. Middleton, F.M., Lichtenberg, J.J., & Taft, R.A., 1960. Industrial wastes – measurements of organic contaminants in the nation's rivers. *Industrial & Engineering Chemistry*, 52(6), pp. 99A–102A.
3. US Department of Health, Education, and Welfare – Public Health Service, 1959. Report on the recovery of orthonitrochlorobenzene from the Mississippi River. June 22.
4. Howard, P.H., Santodonato, J., Saxena, J., Malling, J., & Greninger, D., 1976. *Investigation of Selected Potential Environmental Contaminants: Nitroaromatics*. US EPA, Office of Toxic Substances. EPA-560/2-76-010.

5. Bucher, J.R., 1993. NTP technical report on toxicity studies of 2-chloronitrobenzene and 4-chloronitrobenzene (CAS Nos. 88-73-3 and 100-00-5). NIH publication 93-3382. US Department of Health and Human Services, National Institutes of Health, 1993. Available at: http://ntp.niehs.nih.gov/ntp/htdocs/st_rpts/tox33.pdf. p. 19.

6. US EPA, 2014. United States Environmental Protection Agency Regions 3, 6, and 9, 2014. Regional Screening Levels for Chemical Contaminants at Superfund Sites. Available at: www.epa.gov/reg3hwmd/risk/human/rb-concentration_table/index.htm (accessed September 14, 2014).

7. Council on Environmental Quality (CEQ), 1971. Toxic Substances, reprinted in Staff House Committee on Interstate and Foreign Commerce, 94th Congress, 2nd Session, Legislative History of the Toxic Substances Control Act (TSCA Legislative History) at 760 (Comm. Print, 1976), p. 776.

8. Illinois Department of Public Health, 2009. Public Health Assessment, Sauget Area 2 Landfill Sites P, Q and R, Sauget, St. Clair County, Illinois. EPA Facility ID # ILD000672329. Prepared on behalf of the Agency for Toxic Substances and Disease Registry. Available at: www.atsdr.cdc.gov/HAC/pha/pha.asp?docid=605&pg=1 (accessed May 2014).

9. US EPA, 2018. ChemView: Chemical Data Reporting, 2-Chloro-1-nitrobenzene (CAS 88-73). Data last updated on December 29, 2018. Available at: https://chem view.epa.gov/chemview (accessed January 1, 2019).

10. Sellers, K., 2015. *Product Stewardship, Life Cycle Analysis and the Environment.* Section 4.1.4. Boca Raton: Taylor & Francis/CRC Press.

11. 33 U.S. Code §407 – Deposit of refuse in navigable waters generally.

12. Rotman, M., 2010. Cuyahoga River Fire. Cleveland Historical. Web page last updated April 27, 2017. Available at: https://clevelandhistorical.org/items/show/63 (accessed February 15, 2019).

13. Ruckelshaus, W.D., 1988. Environmental regulation: The early days at EPA. *EPA Journal*, 14(2), pp. 4–5.

14. Lewis, J., 1988. Looking backward: A historical perspective on environmental regulations. *EPA Journal*, 14(2), pp. 26–29.

15. Federal Water Pollution Control Act (the "Clean Water Act") 33 U.S.C. §§1251-1387.

16. US EPA, 2018. Toxic and Priority Pollutants Under the Clean Water Act. Web page last updated November 30, 2018. Available at: www.epa.gov/eg/toxic-and-priority-pollutants-under-clean-water-act (accessed February 17, 2019).

17. 40 CFR §117.3 Determination of reportable quantities. Available at: www.ecfr.gov /cgi-bin/text-idx?node=se40.22.117_13&rgn=div8 (accessed February 17, 2019).

18. 42 U.S.C. §§ 7401 et seq. Enacted as: the "Clean Air Act", on December 17, 1963.

19. US EPA, 2017. Initial List of Hazardous Air Pollutants with Modifications. Web page last modified March 16, 2017. Available at: www.epa.gov/haps/initial-list-hazardous-air-pollutants-modifications (accessed February 17, 2019).

20. 42 U.S.C. §6901 et seq. (1976).

21. 15 U.S.C. §2601 et seq. Toxic Substances Control Act.

22. 42 U.S.C. §300 et seq. Subchapter XII – safety of public water systems.

23. US EPA, 2018. National Primary Drinking Water Regulations. Web page last updated December 18, 2018. Available at: www.epa.gov/ground-water-and-drinking-water/national-primary-drinking-water-regulations (accessed February 17, 2019).

24. US EPA, 2017. Secondary Drinking Water Standards: Guidance for Nuisance Chemicals. Web page last updated March 8, 2017. Available at: www.epa.gov/dwstan

dardsregulations/secondary-drinking-water-standards-guidance-nuisance-chemicals (accessed February 17, 2019).

25. 42 U.S.C. §§9601-9675. Title I – hazardous substances releases, liability, compensation. Sec. 101.

26. Brown, T.G., III, 1984. Superfund and the national contingency play: How dirty is dirty-how clean is clean. *Ecology LQ*, 12, p. 89.

27. US EPA, 1992. Report of the EPA Hazardous Substances Task Force. Office of Emergency Response. April.

28. US EPA, 1992. Progress Toward Implementing Superfund. Fiscal Year 1992 Report to Congress. EPA 540-R-95-145; PB96-963210.

29. US EPA, 1993. Progress Toward Implementing Superfund. Fiscal Year 1993 Report to Congress. EPA 540-R-95-146; PB96-963211.

30. US EPA, 1996. Hazardous Waste Characteristics Scoping Study. EPA 530-R-96-053, November 1996.

31. Strain, H.H., & Sherma, J., 1967. Michael Tswett's contributions to sixty years of chromatography. *Journal of Chemical Education*, 44(4), p. 235.

32. Bartle, K.D., & Myers, P., 2002. History of gas chromatography. *TrAC Trends in Analytical Chemistry*, 21(9–10), pp. 547–557.

33. US EPA, 2007. Region III Quality Assurance Team. TIC Frequently Asked Questions, Revision No. 2.5. Tentatively Identified Compounds. What are they and why are they important? Available at: www.epa.gov/sites/production/files/2015-06/documents/tics.pdf (accessed June 16, 2019).

34. Shimadzu, 2015. Shimadzu Guide to US EPA Method 8260 for Analysis of Volatile Organic Compounds in Ground Water and Solid Waste. Application News. No. GCMS-1503.

35. US EPA, 1998. Method 8270D. Semivolatile Organic Compounds by Gas Chromatography/Mass Spectrometry (GC/MS). Available at: www.epa.gov/sites/production/files/2015-07/documents/epa-8270d.pdf. Retrieved on April 25, 2019.

36. Marchal, R., Penet, S., Solano-Serena, F., & Vandecasteele, J.P., 2003. Gasoline and diesel oil biodegradation. *Oil & Gas Science and Technology*, 58(4), pp. 441–448.

37. Interstate Technology Regulatory Council, 2017. *Per- and Polyfluoroalkyl Substances (PFAS) Fact Sheets: Introduction.*

38. US EPA, 2019. Technical BRIEF: Perfluoroalkyl and Polyfluoroalkyl Substances (PFAS): Methods and guidance for sampling and analyzing water and other environmental media. February 2019 Update.

39. Interstate Technology Regulatory Council, 2018. Site Characterization Considerations, Sampling Precautions, and Laboratory Analytical Methods for Per-and Polyfluoroalkyl Substances (PFAS).

40. US EPA, 2013. What Is the TSCA Chemical Substance Inventory? Web page last updated on April 03, 2013. Available at: www.epa.gov/oppt/newchems/pubs/invntory.htm (accessed June 21, 2013).

41. Massey, R., & Jacobs, M., 2013. United Nations Global Chemicals Outlook – Towards Sound Management of Chemicals. Chapter I: Trends and Indicators. Environment Programme. Available at: www.unep.org/hazardoussubstances/Portals/9/Mainstreaming/GCO/The%20Global%20Chemical%20Outlook_Full%20report_15Feb2013.pdf (accessed July 2013).

42. American Chemistry Council, 2012. *2012 Guide to the Business of Chemistry. Business of Chemistry (Annual Data 2012).*

43. US EPA, 2019. EPA Releases First Major Update to Chemicals List in 40 Years. Press release February 19, 2019. Available at: www.epa.gov/newsreleases/epa-releases-first-major-update-chemicals-list-40-years (accessed May 1, 2019).

44. US EPA, 2012. High Production Volume (HPV) Challenge – Basic Information. Web page last updated April 12, 2012. Available at: www.epa.gov/hpv/pubs/general/basicinfo.htm (accessed July 12, 2013).
45. US EPA, 2004. Status and Future Directions of the High Production Volume Challenge. Available at: www.epa.gov/hpv/pubs/general/hpvreport.pdf (accessed July 12, 2013).
46. US EPA, 2013. High Production Volume Information System (HPVIS) – HPV Chemical Hazard Characterizations. Web page last updated July 12, 2013. Available at: http://iaspub.epa.gov/oppthpv/hpv_hc_characterization.get_report?doctype=2 (accessed July 12, 2013).
47. US EPA, 1997. Chemistry Assistance Manual for Premanufacture Notification Submitters. Chapter 1. EPA 744-R-97-003. March, 1997. Available at: www.epa.gov/oppt/newchems/pubs/chem-pmn/chap1.pdf (accessed June 28, 2013).
48. Anonymous, 2011. Regulatory framework for the management of chemicals (REACH), European Chemicals Agency. Web page last updated February 23, 2011. Available at: http://europa.eu/legislation_summaries/internal_market/single_market_for_goods/chemical_products/l21282_en.htm (accessed November 11, 2013).
49. US EPA, 1976. *Interim Procedures & Guidelines for Health Risk and Economic Impact Assessments of Suspected Carcinogens.* US Environmental Protection Agency. May 9.
50. Albert, R.E., Train, R.E., & Anderson, E., 1977. Rationale developed by the Environmental Protection Agency for the assessment of carcinogenic risks. *Journal of the National Cancer Institute,* 58(5), pp. 1537–1541.
51. US EPA, 1986. Guidelines for carcinogen risk. *Federal Register,* 51(185), p. 33992. September 24, 1986.
52. US EPA, 1985. Chemical carcinogens: A review of the science and its associated principles. *Federal Register,* 50(50), p. 10372. March 14, 1985.
53. US EPA, 1986. Guidelines for the health assessment of suspect developmental toxicants. *Federal Register,* 51(185), p. 34028. September 24, 1986.
54. US EPA, 1991. *Final Guidelines for Developmental Toxicity Risk Assessment.* US Environmental Protection Agency. Risk Assessment Forum. EPA/600/FR-91/001. December.
55. US EPA, 1999. *Guidelines for Carcinogen Risk Assessment.* US Environmental Protection Agency. Risk Assessment Forum. EPA/630/P-03/001B. March.
56. US EPA, 2005. *Guidelines for Carcinogen Risk Assessment.* US Environmental Protection Agency. Risk Assessment Forum. NCEA-F-0644. July.
57. US EPA, 2002. *A Review of the Reference Dose and Reference Concentration Process. Final Report.* US Environmental Protection Agency. Risk Assessment Forum. EPA/630/P-02/002F. December.
58. US EPA, 1995. *New Policy on Evaluating Health Risks to Children [Memorandum].* US Environmental Protection Agency. October 20.
59. US EPA, 2013. *Reaffirmation of the US Environmental Protection Agency's 1995 Policy on Evaluating Health Risks to Children [Memorandum].* US Environmental Protection Agency. October 31.
60. US EPA, 1976. *Quality Criteria for Water.* US Environmental Protection Agency. EPA 440-9-76-023.
61. US EPA, 1992. *Framework for Ecological Risk Assessment.* US Environmental Protection Agency. Risk Assessment Forum. EPA/630/R-92/001. February.
62. US EPA, 1991. *Summary Report on Issues in Ecological Risk Assessment.* US Environmental Protection Agency. Risk Assessment Forum. EPA/625/3-91/018. February.
63. US EPA, 1998. *Guidelines for Ecological Risk Assessment.* US Environmental Protection Agency. Risk Assessment Forum. EPA/630/R-95/002F. April.

64. US EPA, 2014. Pesticides: Consideration of volatilization in pesticide risk assessment. *Federal Register*, 79(58), p. 16791. March 26, 2014.

65. US EPA, 2012. *Standard Operating Procedure for Residential Pesticide Exposure Assessment*. US Environmental Protection Agency, Office of Pesticide Programs, Health Effects Division. October.

66. US EPA, 2013. *Part 158 Toxicology Data Requirements: Guidance for Neurotoxicity Battery, Subchronic Inhalation, Subchronic Dermal and Immunotoxicity Studies*. US Environmental Protection Agency, Office of Pesticide Programs. May.

67. US EPA, 2016. *Process for Evaluating and Implementing Alternative Approaches to Traditional In Vivo Acute Toxicity Studies for FIFRA Regulatory Use*. US Environmental Protection Agency, Office of Pesticide Programs. February.

68. US EPA, 2012. *Benchmark Dose Technical Guidance*. US Environmental Protection Agency. Risk Assessment Forum. EPA/100/R-12/001. June.

69. OECD, undated. OECD Test Guidelines for the Chemicals. Available at: www.oecd. org/env/ehs/testing/oecdguidelinesforthetestingofchemicals.htm (accessed June 7, 2019).

70. A. Bergman, J.J. Heindel, S. Jobling, K.A. Kidd, and R.T. ZoellerEditors, 2012. *State of the Science of Endocrine Disrupting Chemicals – 2012*. Inter-Organisation Programme for the Sound Management of Chemicals (IOMC).

71. IPCS, 2002. *Global Assessment of the State of the Science of Endocrine Disruptors*. T. Damstra, S. Barlow, A. Bergman, R. Kavlock, and G. Van Der Kraak eds. International Programme on Chemical Safety. WHO/PCS/EDC/02.2.

72. US EPA, 2018. Basic Information about the Integrated Risk Information System. Web page last updated October 18, 2018. Available at: www.epa.gov/iris/basic-information-about-integrated-risk-information-system#history (accessed January 1, 2019).

73. US EPA, IRIS Assessments, List A-Z. Exported from https://cfpub.epa.gov/ncea/iris_drafts/atoz.cfm?list_type=alpha (accessed January 11, 2019).

74. US EPA, 1987. 1,2-Dichloroethane; CASRN 107-06-2. Integrated Risk Information System (IRIS) Chemical Assessment Summary. Available at: https://cfpub.epa.gov/ncea/iris/iris_documents/documents/subst/0149_summary.pdf (accessed February 14, 2019).

75. ECHA, 2019. 1,2-Dichloroethane. https://echa.europa.eu/registration-dossier/-/registered-dossier/15430/1.

76. National Academy of Sciences, 2011. Review of the Environmental Protection Agency's Draft IRIS Assessment of Formaldehyde. p. 4. Available at: http://nap.edu/13142 (accessed February 14, 2019).

77. National Academies of Sciences, Engineering, and Medicine, 2018. *Progress Toward Transforming the Integrated Risk Information System (IRIS) Program: A 2018 Evaluation*. Washington, DC: The National Academies Press. Available at. https://doi.org/10.17226/25086 (accessed April 16, 2019).

78. OSHA, undated. Permissible Exposure Limits – Annotated Tables. Available at: www.osha.gov/dsg/annotated-pels/(accessed April 16, 2019).

79. US EPA, 2016. National Recommended Water Quality Criteria. Web site last updated December 13, 2016. Available at: https://19january2017snapshot.epa.gov/wqc/national-recommended-water-quality-criteria_.html (accessed April 16, 2019).

80. European Chemicals Agency, 2019. Registered Substances. Available at: https://echa.europa.eu/information-on-chemicals/registered-substances (accessed June 23, 2019).

3 Emergence of "New" Contaminants

What makes a new contaminant emerge as a pressing environmental issue? History shows that a new understanding of the extent of exposure, coupled with fresh recognition of possible hazards and catalyzed by public scrutiny and outrage, results in calls for action. This chapter describes the factors that can result in the emergence of a "new" contaminant, illustrating each factor with examples and providing the reader with an analytical framework for assessing and managing risks.

3.1 IDENTIFYING POTENTIAL FOR EXPOSURE

As the story of orthonitrochlorobenzene in Chapter 2 illustrated, contaminants often emerge when data show that a chemical has been released into the environment. This section of the book discusses the potential for analysts to identify exposures, considering three factors:

- The universe of chemicals in commerce
- The potential for release, given the nature of their manufacture, storage, and use
- Environmental and biological monitoring programs.

3.1.1 CHEMICALS IN COMMERCE

Even practitioners who have spent decades investigating and cleaning up hazardous waste sites are often surprised to learn that they have focused on only a handful of the chemicals in use. Consider the numbers below.

The scope of site investigation and remediation is often driven by the availability of cleanup goals, as described in Chapter 2. The United States Environmental Protection Agency's (US EPA) Integrated Risk Information System (IRIS) database provides hazard information on 568 chemicals that can be used to derive cleanup goals [1]. This list can essentially bound the universe of chemicals investigated and remediated at a site. After all, if no benchmark exists to enable a project team to assess the risk that might drive cleanup, then investigation and remediation lack focus. In practice, project teams may focus on even fewer substances than those listed in IRIS. In the Commonwealth of Massachusetts, for example, regulators have developed a list of "Method 1" cleanup levels for about 120 potential groundwater contaminants [2]. Such subsets of chemicals can often define site work.

Approximately 41,000 chemicals are in commerce in the United States [3], approximately 4,300 of which are manufactured or imported at more than 1 million pounds (approximately 450,000 kilograms) per year. Chemicals can be released to the environment through air, water, and waste during manufacture and use; some are also incorporated into products intended for widespread or dispersive use that may result in environmental distribution.

The fact that a chemical is in commerce, even at a high volume, does not mean that it will become a significant environmental contaminant. However, information about the number and mass of chemicals in commerce provides important context for considering the potential for new environmental contaminants to emerge.

In summary, the data reported here show a broad gap between the number of chemicals in commerce and the number of contaminants we look for at sites. That gap reflects several factors:

- As noted above, the fact that a chemical is in commerce, even at a high volume, does not mean that it will be released such that it becomes a significant environmental contaminant. Sections 3.1.2 and 3.1.3 examine the mechanisms for release and transport.
- Our monitoring techniques do not always enable us to identify and quantify chemicals that are not part of routine analyte lists, which are based on lists of acknowledged contaminants. Section 2.3 discussed this point; Sections 3.1.4 and 3.1.5 explore the related point of how we use analytical techniques, some of them newly developed or adapted, to get "early warning" data on emerging contaminants.
- We have not had the toxicological or ecotoxicological data to quantify hazards for many chemicals, and consequently have lacked the ability to assess risk or target cleanup results. Section 3.2 explores this point and recent developments in detail.

3.1.2 POTENTIAL RELEASE TO THE ENVIRONMENT

Where do emerging contaminants originate? The answers to this deceptively simple question can illuminate our ability to predict and recognize an emerging contaminant. Many hazardous waste sites and the emerging contaminants at those sites originated, often decades ago, from uncontrolled releases of chemicals to the environment. However, emerging contaminants and our awareness of those contaminants are not limited to such releases. Everyday activities in the manufacture and use of chemicals can result in permitted or incidental releases of chemical compounds to air and wastewater. Contaminants may first emerge as a concern after they are detected in drinking water samples, sewage effluents, or biological samples, despite being used or present for many years or decades. Monitoring of drinking water supplies sometimes uncovers evidence of a "new" environmental contaminant, as was the case with identification of 1,4-dioxane and per- and polyfluoroalkyl substances (PFAS) chemicals through the UCMR3 (Unregulated Contaminant Monitoring Rule) testing. Finally, emerging contaminants may be discovered through investigation of hazardous waste sites using new analytical methods or by assessing tentatively identified compounds.

Consequently, as we explore the potential for new contaminants to emerge we must consider not only accidental releases, but also the chemicals in commerce and the potential for them to be released during manufacture and use, both historically and currently. It is in these cases, when researchers or regulatory monitoring programs uncover widespread evidence of a "new" contaminant, that those findings – regardless of the source of contamination – have consequences for those working in site investigation and remediation and other aspects of environmental protection.

3.1.2.1 Uncontrolled Releases

This book germinated within the perspective of emerging contaminants and how they may affect sampling and remediation at hazardous waste sites. Such sites have typically originated because of an uncontrolled release to the environment. Under the Comprehensive Environmental Response, Compensation, and Liability Act of 1980 (CERCLA) for example [4]:

> The term "release" means any spilling, leaking, pumping, pouring, emitting, emptying, discharging, injecting, escaping, leaching, dumping, or disposing into the environment (including the abandonment or discarding of barrels, containers, and other closed receptacles containing any hazardous substance or pollutant or contaminant).

The US and many other countries regulate the manufacture, storage and use of chemicals in order to minimize such releases. Such releases can still occur, however. In 2019, the United Nations Environment Programme (UNEP) characterized global releases from chemical accidents as "significant", describing three major types of releases as follows [5]:

- The approximately 3,500 mine tailing impoundments in existence often hold hazardous chemicals. Each year, 2 to 5 major failures and 35 minor failures occur.
- Chemical accidents can be difficult to track based on the lack of reporting regimes in some countries. According to one study of news reports between October 2016 and September 2017, accidents occurred at 667 facilities globally with 184 taking place at chemical processing sites.
- Chemical releases can also result from natural disasters. As reported in the *Houston Chronicle*, for example, more than 100 Hurricane Harvey toxic releases were reported associated with the petrochemical industry in the area [6].

Methyl tert-butyl ether (MTBE), which emerged as a contaminant circa 2000, also illustrates the issues that can emerge as a result of uncontrolled releases. Regulators initially saw MTBE as a solution to air pollution. Adding MTBE to gasoline fulfilled the oxygenate requirements in the 1990 Clean Air Act

Amendments and reduced the emissions of smog-forming pollutants (volatile organic compounds and nitrogen oxides) and certain toxic chemicals (such as benzene) [7]. As reported by the Agency for Toxic Substances and Disease Registry (ATSDR), the market demand anticipated to result from the oxygenate requirements of the 1990 Clean Air Act amendments was approximately 65 million pounds (29.4 million kilograms) per day per day in the United States [8].

MTBE is highly soluble and mobile. When underground storage tanks containing MTBE-enriched gasoline leaked or gasoline spilled, plumes of groundwater contamination spread. Recognition that MTBE could create environmental problems first occurred in 1996 when the city of Santa Monica found MTBE in drinking water supplies at levels as high as 610 µg/L, and shut down two wellfields as a result [9]. At that time ATSDR identified the effects of exposure as potentially including irritation upon inhalation and neurological effects, among others; the Agency also acknowledged limitations to understanding exposures saying, "Because it is not presently considered a major harmful pollutant, it is usually not included in routine national monitoring programs for liquids" [10].

US EPA added MTBE to the UCMR list and monitored drinking water supplies around the country for MTBE between 2001 and 2003. The data showed that public water supplies in 14 states contained detectable MTBE at concentrations up to 49µg/L (average detected value 15.2µg/L and median 9.2µg/L). No advisory level was available at that time to benchmark those data [11] (see Section 3.1.4 for tabulated data).

As of 2007, two dozen states had banned MTBE in gasoline either completely or partially [12]. By the time the use of MTBE in gasoline was phasing out, substantial pollution had occurred. One expert team estimated at that time that cleanup costs would eventually total $2 billion [13].

The manufacture and certain uses of MTBE continue. In 2015, between 1 billion to 5 billion pounds (0.5 billion to 2.3 billion kilograms) of MTBE was manufactured in or imported into the United States [14]. According to Toxic Release Inventory (TRI) data from 2016, on- and off-site disposal or other releases reported by manufacturers totaled 1,743,167 pounds (approximately 790,000 kilograms) [14].

Despite some well-publicized present-day releases, environmental conditions often result from historical or current use and distribution through manufacturing. The next section explores this topic.

3.1.2.2 Release during manufacture, processing and distribution

The US EPA, and corresponding agencies around the world, regulate the release of chemicals during manufacturing, distribution, and storage of chemicals. In the US these mandates include the National Pollutant Discharge Elimination System and Spill Prevention, Control, and Countermeasure regulations authorized under the 1972 Clean Water Act; restrictions on discharge to air under the Clean Air Act; and management of waste under the Resource

Conservation and Recovery Act (RCRA). Other regulatory programs specify the conditions for the transport of hazardous substances.

Discharge permits issued under environmental regulations generally focus on specific compounds that are regulated or under scrutiny. As described in Section 2.2, some of the lists of regulated chemicals are decades old and therefore may not represent the current potential for pollution. Permits may not account for all the possible hazardous chemicals being discharged, may no longer appropriately monitor releases (i.e., monitoring daily limits instead of concentrations), and the discharge limits may be based on outdated hazard data.

The US EPA collects data on chemicals released to the environment in the US during routine manufacturing operations and maintains those data in the TRI as part of the Emergency Planning and Community Right to Know Act program established in 1986. This program covers 695 chemicals listed under 33 chemical categories, including those that can cause (1) cancer or other chronic human health effects, (2) significant adverse acute human health effects, or (3) significant adverse environmental effects [15]. The US EPA maintains a publicly available database tracking the management of these chemicals, and each year publishes a TRI National Analysis report that provides a summary of the reported TRI data, evaluates data trends and provides an interpretation of the findings.

TRI data from 2007 to 2017 indicate that the total mass discharge trend of chemicals through air emissions, surface water discharge, or land disposal appears approximately stable over the past 10 years (see Figure 3.1) based on the reporting data from over 20,000 facilities [16]. These discharges are tracked as "releases" in TRI as they represent chemicals released into the environment, even though they may be permitted discharges with approval by the appropriate regulatory authority.

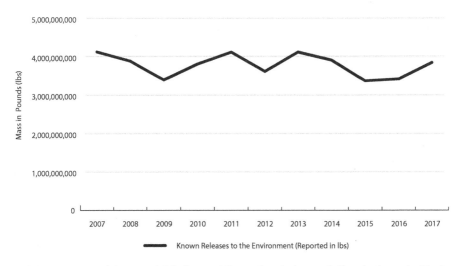

FIGURE 3.1   Total Reported Discharge, Disposal or Release of Chemicals to the Environment over Time as Tracked by the TRI

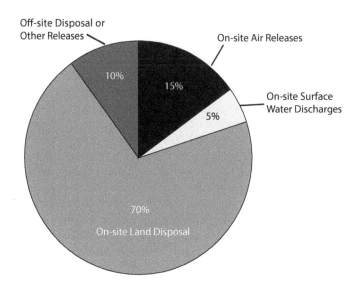

FIGURE 3.2 Distribution of Releases into the Environment by Affected Media, as Tracked by the TRI. Data from 2017

As shown in Figures 3.1 and 3.2, the total mass of chemicals being released into air, surface water, or via land disposal has been in excess of 3 billion pounds (1.4 billion kilograms) per year in the United States over the past 10 years. In 2017, approximately 3.8 billion pounds (1.7 billion kilograms) of TRI reported chemicals were released to the environment, of which approximately 15% (580 million pounds or 263 million kilograms) were emitted into the air and 5% (190 million pounds or 86 million kilograms) were discharged to surface water. On-site land disposal accounted for the largest mass, and includes categories such as underground injections (200 million pounds or 91 million kilograms) and disposal into RCRA Subtitle C (hazardous waste) landfills (88 million pounds or 40 million kilograms). While these are permitted or regulated releases, the volume of chemicals being disposed of each year is significant.

In Section 3.1.3 we'll look briefly at the mechanisms which can spread releases to air and water around the world, with the potential for contaminants to emerge as concerns far from the point of discharge. Of greatest concern in this respect are the persistent, bioaccumulative and toxic (PBT) chemicals released to air and water. To assess the potential for contaminants to emerge from discharges tracked under TRI, several data sets were compared: TRI chemicals that are also on the High Production Volume (HPV) chemical list were compared to the list of PBT chemicals and the list of Substances of Very High Concern (SVHC) chemicals. The compounds on three lists may elicit the greatest concerns with respect to contamination local to the source and to global transport. Of the 299 TRI HPV chemicals, 33 are SVHC and 5 are PBT.

Just as important as understanding what TRI data *do* tell us about such compounds is understanding what the TRI data *don't* tell us. The TRI data described above pertain to the emissions of approximately 695 chemicals out of the 40,655 chemicals in commerce in the United States (of which approximately 4,300 are or have been produced or imported at 1 million pounds per year or more). Not all of those chemicals are released to the environment in significant quantities, of course. But consider the following comparisons, intended to answer the question: *"Which of the chemicals manufactured in large quantities and that may present a significant hazard are not tracked under TRI?"* The list of HPV chemicals was compared to the list of PBT chemicals to identify those of potentially highest concern. Comparison of the chemicals on both lists to the chemicals tracked under TRI indicated that 45 HPV/PBT chemicals are not tracked under TRI. In other words, 45 chemicals considered to be persistent in the environment and also produced at more than 1 million pounds per year are not currently tracked under TRI. This result does *not* mean that 45 contaminants will emerge as new concerns. It does indicate that, depending upon the conditions of manufacture, use, and release of those chemicals, one or more might emerge as an environmental issue, particularly if new hazard data show previously unknown consequences from exposure.

3.1.2.3   Release during use

In addition to accidental releases and releases during manufacturing, chemicals may be released as part of their use. This includes, for example, chemicals in products intentionally designed to be sprayed on the ground; chemicals in commercial products that may be washed down the drain during or use; or chemicals in consumer products that can be released to the environment after use and disposal (i.e., landfilling). Like many of the release mechanisms discussed earlier, these releases often occur under permit or within the bounds of the environmental and product laws in force at the time. Also, as for the release mechanisms discussed previously, data on release quantities may be limited for many chemicals.

Specific examples of chemicals released to the environment during use are plentiful. Most recently, the chemicals present in aqueous film forming foam (AFFF) have come under increased scrutiny as these products were specifically designed to be sprayed on the ground, buildings, or other equipment that was on fire. Annual use of AFFF released to soil and water was estimated to be 5 to 10 tons per year from 1965 to 1974 and 50 to 100 tons globally through use in fire-fighting [17]. As discussed in this book, these applications have led to the detection of AFFF chemicals – i.e., perfluorooctanoate (PFOA) and perfluorooctane sulfonate (PFOS) – in municipal and other water supplies. Data on PFAS also indicate that these chemicals are present in landfills associated with consumer products such as microwave popcorn bags, food wrappers, nonstick cookware, and dental floss, to name just a few. In an evaluation of 18 landfills across different climatic zones and ages, PFOA and PFOS were detected in landfill leachate at average concentrations ranging from 0.1ug/L to 1ug/L and

0.003ug/L to 0.2ug/L, respectively [18]. These values compare to the US EPA Health Advisory of 0.07ug/L, suggesting that disposal of these chemicals after use can also lead to chemical detections above regulatory guidelines.

In other cases, use and then release of a chemical may be less direct. For example, perchlorate use associated with fireworks can lead to release of the chemical to the environment. ATSDR's *Toxicological Profile for Perchlorates* reports that water samples in a lake near a fireworks display in Oklahoma had an increased perchlorate concentration from 0.043ug/L prior to the event to 44.2ug/L after the event [19]. While this type of release may be less common or limited to certain times of the year, it does demonstrate that chemicals, and often emerging contaminants, are present in products and may be released during use. Other outdoor uses of perchlorate sometimes occurred over large areas. Those uses, in fertilizer and explosives among others, also contributed to environmental contamination.

As these examples show, the widespread use of a chemical in an outdoor setting can lead a contaminant to emerge. Whether the concern over that pollution is local or the releases have global effects depends upon the mechanisms for transport.

3.1.3 Global Transport after Release[1]

At a hazardous waste site, project teams typically consider local transport through groundwater, soil, vapors, and surface water. But some contaminants emerge as concerns as a result of transport on a far larger scale. The global mechanisms for transport by water and air are briefly described below. They provide context for information about the spread of chemicals released to the environment during manufacture (as described by TRI data, for example) or use.

3.1.3.1 The Water Cycle and Ocean Currents

Water cycles continually through our environment. The movement of water within a region or more broadly within the world's oceans can carry contamination far from its point of origin, as we'll see in an example to follow.

Water evaporating from the earth's surface falls back to the ground as precipitation. Rainfall and snowmelt can soak into the ground, to be intercepted by the roots of plants or to remain in the earth as groundwater until the flowing groundwater discharges to surface water or is pumped to the surface from a well. Water on the ground surface can run off into surface water, perhaps into a stream that conveys water to a larger river and then to the ocean.

Once in the ocean, the tides and currents driven by wind control the circulation of shallow water. Seven major currents, shown on Figure 3.3[2], move water around the globe: the West Wind Drift (or the Antarctic Circumpolar Current), East Wind Drift, the North and South Equatorial currents, the Peru Current, the Kuroshio Current, and the Gulf Stream. These currents can move quickly. The Gulf Stream, for example, usually travels at a speed of 3 or 4 knots, which is equivalent to 3.5 to 4.6 miles per hour (5.6 to 7.4 kilometers

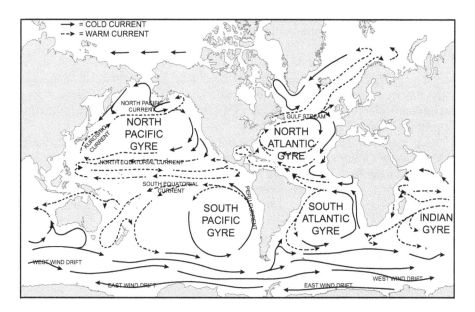

FIGURE 3.3 Ocean Currents and Gyres

per hour). As these currents spiral through the ocean they form five major gyres: the North Atlantic, South Atlantic, North Pacific, South Pacific, and Indian Ocean gyres [20].

Deeper yet, water may enter the "Great Ocean Conveyor Belt" where thermohaline circulation driven by variations in temperature and salinity effects the three-dimensional movement of the ocean's waters around the Earth [21,22]. As Arctic air chills the oceans near Iceland and sea ice forms, the salinity and density of the water at the surface increase until the cold water sinks. This mass, known as the North Atlantic Deep Water, flows slowly to the south deep in the Atlantic Ocean. Near the coast of Antarctica, this mass mixes with water from the Antarctic and deep water circulating from the Pacific and Indian Oceans. Flowing to the east around Antarctica the mass cools, sinking further, and then splits in two. One stream moves north into the Indian Ocean and the other farther east into the Pacific Ocean. As these streams move north and east the water begins to warm, becoming less dense and upwelling toward the surface. Then the currents loop to the south and the west again into the South Atlantic. Water then flows north to the North Atlantic where it enters the conveyor belt again.

Water moves slowly through the Great Ocean Conveyor Belt. According to one estimate, it takes any given cubic meter of water in the ocean approximately 1,000 years to complete the journey. Slow as it is, the sheer volume of water moving through this conveyor – estimated at more than 100 times the flow of the Amazon River [23] – dramatically affects the global ocean nutrient and carbon dioxide cycles. And as we will explore in an example below, this global cycling has the potential to carry chemical contaminants far from their point of origin.

Human activities can release chemical substances at several points in the water cycle. Leachate from an unlined landfill or sewage treated in a septic tank may infiltrate into groundwater. Treated wastewater commonly discharges to surface water. And air pollutants enter the water cycle when they dissolve in precipitation and return to earth.

Each step in the water cycle has the potential to convey chemical substances through the environment by advection, until contaminants degrade. Precipitation can wash chemical substances out of the air and carry them to soil or surface water. On land, chemicals may then re-evaporate or infiltrate to groundwater, or run off to surface water either sorbed to particulates or dissolved in water. Once in surface water, compounds may evaporate or be carried far downstream with flowing surface water, even to remote marine environments.

One research team questioned how to gauge which compounds had the highest potential for long-range transport in water [24]. They began with the criterion under the Stockholm Convention that a chemical may have the potential for long-range transport by air if its half-life in air is greater than two days, and scaled that factor to apply it to transport by water based on the relative velocities of long-range air and water flows. Their calculations suggest that in a river system a substance with a half-life between 8 and 15 days could travel a distance of 700 kilometers (435 miles). In the ocean, considering the speed of Arctic currents, a half-life of 40 to 130 days would allow for a similar transport distance. Chemical compounds with half-lives greater than these calculated values could pose a potential for long-range transport in water. While these screening criteria are not absolute, they do give a sense of the possibility that the flow of water may transport even degradable substances long distances.

Consider, for illustration, the fate and transport of certain perfluorinated compounds. PFOS is one of the most studied compounds within the large diverse group of PFAS chemicals. As discussed above, PFOS and related compounds have had many uses, including electric and electronic parts, firefighting foam, photo imaging, hydraulic fluids, and textile treatments (e.g., waterproofing or stain repellency), although some uses have been phased out [25]. The strong carbon–fluorine bonds in PFOS resist degradation. Hydrolysis tests showed a half-life of approximately 41 years; the indirect photolytic half-life of PFOS has been estimated to be greater than 3.7 years; and, as reported by UNEP, repeated tests of biodegradability have failed to demonstrate any signs of biodegradation [25]. UNEP also reports that PFOS is highly bioaccumulative and can biomagnify, albeit not by the "typical" mechanism of partitioning into fatty tissues. Instead PFOS bioaccumulates by binding to plasma proteins and to proteins in the liver [25].

PFOS may enter domestic sewage as a result of washing treated fabrics or cleaning carpets, for example. Industrial sewage may also contain PFAS compounds from the use of industrial additives, surfactants, and coatings. Conventional wastewater treatment would not degrade PFOS, although it might degrade related compounds to form PFOS. PFOS could sorb to the solids in the wastewater treatment plant or discharge to surface water in the effluent. A literature survey in 2014 [26] found that PFOS had been detected in the effluent of wastewater treatment plants in North America, Europe, and Asia Pacific at concentrations up to 432ng/L.

Such wastewater effluents commonly discharge to surface waters, which ultimately discharge to the ocean. Various research teams have quantified PFOS in rivers and seas; the most sweeping studies have examined the distribution of PFOS in ocean waters. Consider this illustration, which reflects the consequences of wastewater discharges to the River Elbe in Germany.

Effluent from 7 of 9 wastewater treatment plants contained detectable PFOS, at concentrations between 0.5 and 82.2ng/L. Surface water samples from the River Elbe, which reflected the contribution from municipal wastewater effluents and from industrial wastewater effluents or from surface runoff, all contained detectable PFOS, at levels between 0.5 and 2.9ng/L [27]. Samples from the German coast near the discharge of the River Elbe to the North Sea contained 4.09 to 6.06ng/L PFOS; 16 water samples from the North Sea along the German coast contained up to 2.26ng/L PFOS; and three samples from the open North Sea contained up to 0.07ng/L PFOS. (Not all of the samples collected off the German coast or in the North Sea contained detectable PFOS.) [28]. The concentrations were highest close to industrial or highly populated areas but decreased rapidly in open sea water. Sorption to particulates, based on data not reported here, would account for only a fraction of the PFOS mass. A separate research team found similar results, detecting 8ng/L PFOS in water samples collected from the mouth of the River Elbe and 1.8ng/L PFOS in samples from coastal stations [29].

Researchers have ventured even farther asea to measure the levels of PFOS and related compounds in the waters of the world's oceans. They have detected PFOS throughout the ocean to depths of 1000m (3,280 feet) or more. Samples from coastal areas off China, England, Germany, Hong Kong, Japan, and Korea have contained PFOS. Farther out at sea, researchers have detected PFOS in samples from throughout the Pacific Ocean and Atlantic Ocean, and from the Southern (Antarctic) Ocean and Arctic Ocean. PFOS have also been detected in the Greenland Sea, Labrador Sea, Indian Ocean, and the South China Sea. The detected concentrations are generally on the order of tens of picograms per liter (pg/L; 1pg/L = 0.001ng/L) though concentrations of up to 291pg/L have been detected in the open ocean [30–34].

PFOS has also moved deeper in the oceans. In one study between 2002 and 2006, a team of researchers collected samples from 62 locations offshore of Japan and Angola and in the open ocean at various depths [34]. These open-water sampling locations were in the North Atlantic Ocean, Mid-Atlantic Ocean, central to western Pacific Ocean, south Pacific Ocean, South China Sea, Japan Sea, and Indian Ocean. They concluded that the distributions of PFOS and the related compound PFOA, laterally and vertically, were consistent with transport via the Great Ocean Conveyor Belt. The relatively low-density distribution of samples and the relatively short period of use of PFOS (about 50 years) make it difficult to reach a definitive conclusion, but the results suggest, in the researchers' view, the potential for global three-dimensional transport of persistent compounds.

3.1.3.2   Atmospheric Transport

The global mechanisms for atmospheric transport of contamination reflect the movement of winds through the Earth's atmosphere. Meteorologists represent the Earth's atmosphere as a series of concentric layers. Closest to the Earth is the troposphere. This layer, which varies between 8 and 18km (5 and 11 miles) thick at different points on the earth, comprises the planetary boundary layer (i.e., the 1 to 2 kilometers (0.6 and 1.2 miles) closest to the Earth) and the free troposphere. The tropopause is the outer boundary of the troposphere. It separates the troposphere from the next layer, known as the stratosphere.

Air circulates within the troposphere from the equator toward the two poles. In simple terms, the hot air at the equator rises and moves toward the poles; when it encounters cold temperatures at the poles, it becomes more dense and sinks, then circulates back toward the equator.

Near the surface of the Earth, north–south wind flow can be of the same magnitude as east–west flow; in the middle and upper troposphere, west to east flow predominates. Wind speeds in the troposphere generally increase with altitude and are typically strongest in the winter months. The net effect is that winds in the middle-latitude troposphere generally blow from west to east and consequently most intercontinental pollutant transport in the troposphere is from west to east [35].

Jet streams of fast-moving air ride along the tropopause. While these winds typically blow at 129 to 225 kilometers per hour (80 to 140 miles per hour), they can reach speeds of more than 443 kilometers per hour (275 miles per hour) [36].

Two primary mechanisms account for pollutant transport. The first tends to act on a regional scale; the second, global.

In the first mechanism, air pollutants emitted to the planetary boundary layer (PBL) can mix into background conditions and move by advective transport – that is, with the wind – throughout a region. The degree of mixing depends on the turbulence that results from temperature gradients [35]. Air pollutants in the PBL may rise higher into the troposphere or may be carried back to the surface of the Earth with temperature-driven air flow. Some pollutants may sorb to solids; precipitation may return those solids or dissolved contaminants back to the ground. Chemical and photochemical reactions may degrade some substances.

The second pollutant transport mechanism can act more globally. Wind currents can lift localized pollutants up out of the PBL and into the free troposphere. The meteorological conditions off the east coasts of North America and Asia, in particular, can drive this form of transport. Once in the free troposphere, strong winds like the jet streams can move pollutants for long distances without significant dilution or removal. The net effect is this [35]:

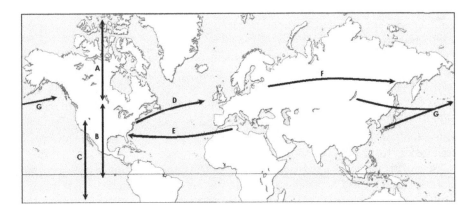

FIGURE 3.4 Major Atmospheric Transport Pathways and Transport Time Scales. [A] Midlatitudes–Arctic Exchange (1–4 weeks). [B] Midlatitudes–Tropics Exchange (1–2 months). [C] Northern Hemisphere–Southern Hemisphere Exchange (approx. 1 year). [D] North America to Western Europe (3–13 days). [E] Eastern Europe to Asia (1–2 weeks). [F] Eastern Europe to Asia (1–2 weeks). [G] Eastern Asia to North America (4–17 days)

Any air pollutant with an atmospheric lifetime of at least three to four days may be transported across most of a continent, a week or two may get it across the ocean, a month or two can send it around the hemisphere, and a year or two may deliver it anywhere on Earth.

Figure 3.4[3] [35] shows how atmospheric transport pathways can carry airborne pollutants around the globe. Such global transport may not occur in a single step or stage. Scientists have coined the terms "global distillation" and the whimsically named "grasshopper effect" to describe a hypothesis that explains how the long-range transport of air pollutants may happen in stages. Global distillation, as the name implies, refers to the sequential volatilization and condensation of a substance. A pollutant released to the atmosphere migrates with the winds until it meets cooler temperatures that change the state of the pollutant and it returns to earth to sorb to soil, overwinter in snow, or dissolve in surface water. When the temperature rises again, perhaps during the next summer season, the compound vaporizes again and can migrate with the prevailing winds. A series of distillation steps can gradually carry a pollutant far from its point of origin, "hopping" across the globe until it reaches steady state in the cold regions of the Arctic or Antarctic. Just as in a distillation column or chromatographic column, this process can effectively separate (or fractionate) substances according to their physicochemical characteristics. More volatile compounds may travel farther than their less-volatile cousins. Persistent semi-volatile substances, depending on environmental conditions such as temperature, precipitation, bioturbation, and the organic content of the soils with which they come into contact, are of particular concern with respect to the potential for long-range transport [37].

While some evidence supports the global distillation hypothesis, other scientists have proposed competing explanations for global transport. The differential removal hypothesis, for example, focuses on the fact that even persistent compounds have some potential to degrade. The originators of this hypothesis postulated that fractionation results not from temperature-driven "distillation" but from different loss rates from the atmosphere that manifest along a gradient of remoteness from emission sources [38].

Scientists have amassed data that demonstrate that long-range transport occurs, notably in data from the Arctic region far from industrial sources of pollution. PFOS, discussed above with respect to long-range transport in water, is one such compound. Scientists have found that PFOS carried by ocean currents and, to a lesser extent air transport [17], have contaminated the Arctic. When the UNEP identified the PFOS family of compounds as Persistent Organic Pollutants under the Stockholm Convention in 2006, scientists recognized that the transport of PFOS could have global consequences, saying [25]:

> Most notable and alarming are the high concentrations of PFOS that have been found in Arctic animals, far from anthropogenic sources. PFOS has been detected in higher trophic level biota and predators such as fish, piscivorous birds, mink, and Arctic biota. Also, predator species, such as eagles, have been shown to accumulate higher PFOS concentrations than birds from lower trophic levels. Even with reductions in manufacturing of PFOS by some manufacturers, wildlife, such as birds, can continue to be exposed to persistent and bioaccumulative substances such as PFOS simply by virtue of its persistence and long-term accumulation.

Concerns over Arctic contamination are not limited to PFOS. Since 1991 the Arctic Monitoring and Assessment Programme (AMAP) has monitored contamination in environmental and biological samples from the Arctic region. Based upon their data, AMAP warns about the potential consequences of both well-known and emerging contaminants. A 2017 report [39] sounded a clear alarm:

> Tens of thousands of chemicals are presently on the market and new substances continue to enter commerce each year. Many of the chemicals currently registered for use have characteristics similar to legacy pollutants, including a potential to reach the Arctic; however, most are not subject to international (global) regulation....

> Improved analytical technologies, research and screening programmes continue to reveal the presence of chemicals that have previously gone unnoticed, or were not expected to be present in the Arctic. Although newly detected in the Arctic, these so-called "chemicals of emerging concern", have often been in use and present in the environment for years, even decades. Chemicals found in the Arctic may

originate from local sources within the region or come from distant locations. The detection of a new substance in the Arctic that has no local sources is particularly important, as it provides evidence of the chemical's potential to disperse globally. As new substances and their breakdown products continue to be discovered, the notion of what constitutes an "environmental pollutant" warranting concern also changes, and updated regulatory actions may be needed.

Based on recent monitoring data, AMAP has identified the following "chemicals of emerging Arctic concern":

- Brominated flame retardants
- Chlorinated flame retardants
- Chlorinated paraffins
- Current-use pesticides
- Halogenated natural products
- Hexachlorobutadiene
- Organophosphate-based flame retardants
- Organotins
- Pentachlorophenol
- Per- and polyfluoroalkyl substances
- Pharmaceuticals and personal care products
- Phthalates
- Plastics and microplastics
- Polychlorinated naphthalenes
- Polycyclic aromatic hydrocarbons
- Siloxanes
- Unintentionally generated polychlorinated biphenyls.

Some of these contaminants are relatively well known, such as polycyclic aromatic hydrocarbons; others are not. We'll return to examining chemicals of emerging Arctic concern in Section 3.3 of this book. In the next sections we'll explore other programs besides AMAP which collect and analyze environmental and biological data that may give evidence of emerging contaminants.

3.1.4 ENVIRONMENTAL MONITORING

It is common sense, and illustrated by the AMAP work described in the preceding section, that contaminants emerge only after we become aware of the potential for exposure. Consider, for example, how perchlorate emerged as an environmental contaminant, as illustrated in Figure 3.5 and described below [40–46].

The first widespread releases of perchlorate to the environment occurred nearly two centuries ago. American farmers have used Chilean nitrate fertilizer, the largest known natural source of perchlorate, since 1830. Its use peaked in the years 1909 to 1929 at a total of 19 million tons, or nearly one million tons per year [47]. Perchlorate was first used commercially in World War II in rockets and missiles and later found use in the manufacture

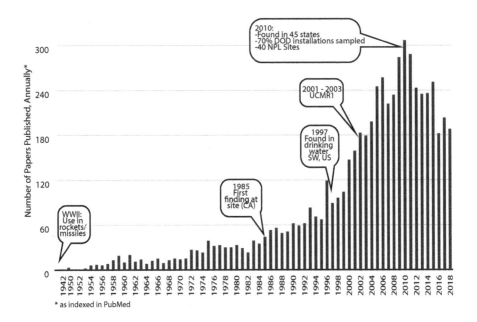

FIGURE 3.5   Emergence of Perchlorate as a Contaminant of Concern

of munitions, explosives, and fireworks [47]. Perchlorates may also have been used in small amounts in certain high-volume consumer products [48].

Perchlorate was not recognized as an environmental contaminant until it was detected in the environment in 1985. At that time, chemists did not routinely monitor for perchlorate, and the available test methods had relatively high detection levels that limited the ability to recognize contamination in environmental samples [42].

Interest grew a decade later after chemists detected perchlorate in drinking water supplies in the southeastern United States. Initial findings of perchlorate in environmental samples and concerns over the potential health effects of exposure drove the development of more sensitive analytical methods. Concerns over perchlorate accelerated as teams using the new, more sensitive test methods found perchlorate in an increasing number of samples [42].

As concerns increased the US EPA added perchlorate to the UCMR. This provision of the Safe Drinking Water Act requires public water supplies to monitor unregulated contaminants. The data collected from this monitoring program, which reflected the analysis of 34,331 samples from 3,865 public water supplies, showed widespread perchlorate contamination [46]:

- Approximately 4% of the public water supplies tested contained perchlorate (in at least 1 entry/sampling point) at concentrations ≥4μg/L. These detections occurred in samples from 26 states and 2 territories.
- Approximately 2% of the samples contained perchlorate at levels ≥4μg/L.

• The average detected concentration of perchlorate was 9.85µg/L and the median concentration was 6.40µg/L.

Scientific interest in the occurrence and health effects of perchlorate grew as awareness of environmental contamination increased. Figure 3.5 illustrates this growing interest with the number of papers published each year on perchlorate. The body of scientific knowledge that resulted fed concerns over perchlorate as a contaminant.

Within a relatively short time after perchlorate emerged as a contaminant, it had been detected in environmental samples from 45 states, the majority of Department of Defense (DOD) sites investigated, and 40 National Priority List sites under CERCLA. In hindsight, that finding is not surprising because of the widespread outdoor use of products containing perchlorate. The emergence of perchlorate as a contaminant can be traced to a series of factors that related primarily to knowledge of exposure: the recognition of the use of perchlorate-containing products; finding perchlorate in multiple drinking water supplies; and the development of analytical techniques that enabled the detection of lower and lower environmental concentrations.

US EPA has been slow to regulate perchlorate as a drinking water contaminant or to set cleanup goals. In 1999 the US EPA published interim guidance which stated that provisional cleanup levels or action levels would range from 4–18µg/L. In January 2009 the Agency issued an Interim Health Advisory for perchlorate of 15µg/L; in 2011, it determined that perchlorate met the criteria for regulation under the Safe Drinking Water Act. The Federal Register notice [49] of this finding stated that:

> [P]erchlorate may have an adverse effect on the health of persons; perchlorate is known to occur or there is a substantial likelihood that perchlorate will occur in public water systems with a frequency and at levels of public health concern; and in the sole judgment of the Administrator, regulation of perchlorate in drinking water systems presents a meaningful opportunity for health risk reduction for persons served by public water systems. Therefore, EPA will initiate the process of proposing a national primary drinking water regulation (NPDWR) for perchlorate.

Despite that finding, as of this writing in June 2019 over two decades after perchlorate emerged as a contaminant, the US EPA has not promulgated a regulation on perchlorate in drinking water.

This case study of perchlorate and the results of AMAP's monitoring program described in the preceding section illustrate how awareness of widespread exposures can occur as a result of a monitoring program carried out by a research organization or governmental agency. The United States government carries out three well-known monitoring programs. (Other federal and state initiatives collect additional data relevant to emerging contaminants.) The United States Geological Survey National Water-Quality Assessment Project has, since 1991, collected contaminant data in streams, rivers, and groundwater [50]. The US Centers for Disease Control and Prevention (CDC) coordinates the National Biomonitoring Program, which gathers data on the exposure of people in the US to environmental chemicals and toxic substances. The CDC tests biological samples (e.g., urine, blood, serum, breast milk, and meconium) for 300 environmental chemicals or their metabolites [51]. To explore the theme of this book, anticipating developments in emerging contaminants, we'll focus on a third program: monitoring under the UCMR.

As described in Chapter 2, the UCMR under the Safe Drinking Water Act gives the US EPA the authority to collect data for contaminants that do not have health-based standards and that the Agency suspects to be present in drinking water. Every five years, US EPA monitors up to 30 contaminants selected from the Contaminant Candidate List (CCL). The Agency describes the process for selecting analytes as follows [52]:

> EPA reviews contaminants that have been evaluated through existing prioritization processes, including previous UCMR contaminants and the CCL. Additional contaminants may be identified based on current research on occurrence and health effect risk factors.

> Chemicals that are not registered for use in the United States, do not have an analytical reference standard, or do not have an analytical method ready for use are generally removed from consideration.

> EPA further prioritizes remaining contaminants based on more extensive health effects evaluations, typically performed by the Office of Water's Office of Science and Technology. The procedures for evaluating health effects were developed to support the ranking of contaminants for future CCLs.

Table 3.1 summarizes the work to date [53].

A 2018 report by the Task Force on Emerging Contaminants of the National Science & Technology Council provides perspective on these efforts under UCMR. This Task Force, which advises the US president, described Federal agencies' ability to assess emerging contaminants in water supplies in programs such as the UCMR studies. Citing a 2017 study of emerging contaminants in treated water supplies, the Task Force noted that 121 different types of unregulated chemicals and microbes have been detected. However, the Task Force noted significant limitations to the ability to accurately monitor emerging contaminants and expressed the concern that a lack of toxicity data limits the ability to understand the consequences of exposure to emerging contaminants [54].

With that caution in mind, let's look at the data from UCMR1. Table 3.2 summarizes those monitoring results (data in μg/L) [55]. These data show that two contaminants, perchlorate and DCPA mono/di-acid degradates, were detected in a similar number of public water supplies, and at concentrations of the same order of magnitude. Perchlorate "emerged" as a contaminant of concern as described above (as did MTBE, as described previously); DCPA

TABLE 3.1

Contaminants Monitored Under UCMR

Monitoring Campaign	Years	Contaminant Groups
UCMR1	2001–2003	2,4-dinitrotoluene; 2,6-dinitrotoluene
		Acetochlor
		DCPA mono-acid degradate, DCPA di-acid degradate
		4,4'-DDE
		EPTC
		Molinate
		MTBE
		Nitrobenzene
		Perchlorate
		Terbacil
UCMR2	2008–2010	Two insecticides
		Five flame retardants
		Three explosives
		Three parent acetanilides
		Six acetanilide degradates
		Six nitrosamines
UCMR3	2013–2015	Seven volatile organic compounds
		1,4-dioxane
		Six metals
		Chlorate
		PFOS/PFOA and related compounds
		Seven hormones
		Two viruses
UCMR 4	2018–2020	Ten cyanotoxins (microsystins)
		Two metals
		Nine pesticides or byproducts
		Three brominated haloacetic acid groups
		Three alcohols and three semi-volatile compounds

mono/di-acid degradates did not. With respect to the themes of this book it is worth a closer look at DCPA mono/di-acid degradates.

Samples from public water supplies in in 24 States and the Territory of Guam contained DCPA degradates, largely in four general regions: California and the western Rocky Mountain States, the Southeast, the Northeast, and the upper Midwest. Detections occurred more frequently in water supplies originating in groundwater than in surface water. The US EPA estimated that the drinking water supplies for 12.4 million people might contain detectable DCPA degradates, albeit at a low levels [11].

TABLE 3.2

Results of UCMR1 (2001–2003)

Contaminant	Maximum Reporting Limit (MRL)	Analyses	Public Water Supplies with Analyses	Detections	Public Water Supplies with Detections	Concentrations Detected (μg/L)		
						min	max	mean
DCPA mono/di-acid degradates	1	34,278	3,882	787	177	1	190	3.5
Molinate	0.9	34,298	3,879	1	1	5.7	5.7	5.7
MTBE	5	34,131	3,877	26	19	5	49	15.2
Nitrobenzene	10	33,937	3,867	2	2	21.6	100	60.8
Perchlorate	4	34,728	3,870	647	160	4	420	9.9
2,4-dinitrotoluene	2	34,126	3,879	1	1	333	333	333
2,6-dinitrotoluene	2	34,126	3,879	0				
4,4'-DDE	0.8	34,370	3,880	1	1	3	3	3
Acetochlor	2	34,274	3,875	0				
EPTC	1	34,297	3,879	0				
Terbacil	2	34,299	3,879	0				
1,2-diphenylhydrazine	0.5	2,354	310	0				
2,4,6-trichlorophenol	1	2,308	304	0				
2,4-dichlorophenol	1	2,308	304	0				
2,4-dinitrophenol	5	2,305	304	0				
2-methyl-phenol	1	2,308	304	0				
Diazinon	0.5	2,354	310	0				
Disulfoton	0.5	2,348	310	0				
Diuron	1	2,360	308	1	1	2.1	2.1	2.1
Fonofos	0.5	2,354	310	0				
Linuron	1	2,357	308	0				
LL-Nitrobenzene	0.5	2,556	349	0				
Prometon	0.5	2,354	310	0				
Terbufos	0.5	2,349	310	0				

The US EPA summarizes concerns about DCPA mono/di-acid degradates as follows [56].

Dimethyl tetrachloroterephthalate (DCPA), a synthetic organic compound (SOC) marketed under the trade name "Dacthal," is a pre-emergent herbicide historically used to control weeds in ornamental turf and plants, strawberries, seeded and trans-planted vegetables, cotton, and field beans. As of 1990, more than 80 percent of its use was for turf, including golf courses and home lawns. Available data indicate that DCPA use declined significantly over the course of the 1990s. On July 27, 2005, in response to concerns about groundwater contamination (especially for one of the DCPA degradates), the registrant voluntarily terminated most uses for products containing DCPA. DCPA is currently registered only for use on sweet potatoes, egg-plant, kale, and turnips.

DCPA is not particularly mobile or persistent in the environment. Biodegradation and volatilization are the primary dissipation routes. Degradation of DCPA forms two breakdown products, the mono-acid degradate (monomethyl tetrachloroter-ephthalate or MTP) and the diacid degradate (tetrachloroterephthalic acid or TPA). The di-acid, which is the major degradate, is mobile and persistent in the field, with a potential to leach into water.

[T]he Agency calculated a health reference level (HRL) of 0.07 mg/L or 70 µg/L for DCPA and used this HRL for TPA and MTP. ... Based on the cancer data for DCPA and evidence that neither TPA nor DCPA is mutagenic, the Agency con-cludes that TPA is unlikely to pose a cancer risk.

The Agency has made a determination not to regulate the DCPA mono-acid degra-date and/or the DCPA di-acid degradate with a national primary drinking water regulation (NPDWR). Because these degradates appear to occur infrequently at health levels of concern in PWSs [Public Water Supplies], the Agency believes that an NPDWR does not present a meaningful opportunity for health risk reduction. While the Agency recognizes that these degradates have been detected in the PWSs moni-tored under the UCMR 1, only 1 PWS detected the degradates at a concentration above the HRL.

This summary by the US EPA indicates that perchlorate and DCPA mono/di-acid degradates are similar in several ways within this context:

- Detected in similar numbers of public water supplies, at concentrations below action levels:
  - Perchlorate detected at a range of 4 to 420µg/L and mean of 9.9µg/L, at a time when the provisional action level was 4–18µg/L (later revised to an Interim Health Advisory for perchlorate of 15µg/l);
  - DCPA degradates detected between 1 and 190µg/L, with a mean 3.5 versus an action level of 70µg/L

- Compounds are mobile and resist degradation
- Products containing perchlorate and DCPA have been in wide dispersive use in agriculture.Despite these similarities, the initial findings of perchlorate in environmental samples spurred broad investigation of groundwater contamination and data regarding DCPA mono/di-acid degradates did not.

Perhaps the perception of perchlorate and DCPA mono/di-acid degradates has differed in part because of the nature of the products and their use. In addition to agricultural use, perchlorate has been released to the environment from rockets and explosives manufactured for and used by the DOD. Those uses have resulted in plumes of groundwater contamination at current and former military bases. Such environmental issues can conjure up a different emotional reaction than might result from considering the use of DCPA on golf courses and crops.

The uses of DCPA in turf management and agriculture have been permitted under the Federal Insecticide, Fungicide, and Rodenticide Act. The herbicide registrant voluntarily terminated the registration of DCPA for most uses in July 2005, in response to concerns about groundwater contamination (especially for one of the degradates), but the substance is still approved for use on some food crops. The United States Geological Survey estimates that in 2016 approximately 5 million pounds (2.3 million kilograms) of DCPA was used on food crops in the states of California, Oregon, Texas, and Washington [57]. The State of California monitors the concentrations of DCPA and degradates in groundwater. In 2017 it found, for example, that three of seven groundwater samples tested contained DCPA degradates, with concentrations ranging from 0.916 µg/l to 101 µg/l. Toxicologists reviewing the data determined that the highest concentration detected (101 µg/l) should be considered a health concern [58].

Outrage over the use of DCPA smolders. The Environmental Working Group noted in 2019 that nearly 60% of kale samples sold in the US were contaminated with residues of DCPA based on 2017 US Department of Agriculture test data [59]. This nongovernmental organization also pointed out that in 1995 (and in contrast to the 2008 US EPA finding cited above), the US EPA determined that DCPA was a possible human carcinogen [60]. In addition, the State of California determined in 2018 that DCPA is possibly carcinogenic [61]. (The authors mention these classifications of carcinogenicity not to adjudicate their validity but to illustrate that information is publicly available to fuel outrage over DCPA as an emerging contaminant.) We'll return to the potential for outrage and its role in the emergence of a contaminant in Section 3.3 of this book.

3.1.5 Potential for Contaminants to Emerge

The contaminants that society has focused on, largely drawn from regulatory lists developed decades ago, do not necessarily reflect more recent manufacturing or environmental conditions. That gap creates the potential for new contaminants to emerge. The sections above describe how knowledge of the manufacture and use of chemicals can be combined with information about possible releases to the environment to gain perspective on the potential for contaminants to emerge. While the data show some "ambiguity and contradiction", as described in Chapter 1 of this book, they can point to the conditions that may cause a contaminant to emerge.

Modern environmental laws limit the potential for the uncontrolled release of hazardous substances in the United States and many other countries. Accidents still happen. Further, permitted releases from manufacturing and use of chemicals discharge billions of pounds/kilograms of pollution each year. Such releases are not well quantified. The TRI program, for example, tracks 695 chemicals listed under 33 chemical categories. But many more chemicals are obviously made or processed in the United States; 45 chemicals considered to be persistent in the environment (PBT) and also produced at more than 1 million pounds per year (HPV) are not tracked under TRI. This result does *not* mean that 45 contaminants will emerge as new concerns; it does mean that there is no line of sight on potential releases or resulting environmental contamination. Releases may also result from dispersive use. As illustrated by PFAS and perchlorate, use of a chemical outdoors in multiple areas can result in contamination that leads to exposure, which causes concerns.

Releases – whether accidental or under permit – can affect environmental conditions near the point of release or can be carried by wind and water far from the original source. A chemical contained within such releases may emerge as a contaminant of concern in monitoring of water supplies, surface water, other environmental media, or biota. Some monitoring programs generate data relatively near to points of release. Others, notably monitoring in the Arctic, identify emerging contaminants far from their point of origin.

Some monitoring programs look for evidence of an emerging contaminant but don't find it. For example, 16 of the chemicals in the UCMR1 program (Table 3.2) simply were not detected in a monitoring program that included thousands of samples from between 300 and 3,882 public water supply systems (depending on the analyte).

Even finding a contaminant during a monitoring program does not necessarily mean that it will emerge as a contaminant of concern. In order to do so it must present a significant risk, real or perceived. Perchlorate emerged as a contaminant at sites after discovery in drinking water. As described above, DCPA mono/di-acid degradates have also been detected in the public water supply and are the degradation products of an herbicide that has been used in broad agricultural settings. Yet DCPA mono/di-acid degradates have not emerged as contaminants of concern. Part of the difference may lie in the levels described as acceptable by regulators and the way in which those evolved over time, which contributed to perceptions of risk. We'll explore other factors that may account for this difference later in the book.

3.2 EMERGING UNDERSTANDING OF HAZARD

Identification of emerging chemicals is based on part on our understanding of the potential human and environmental health effects of the chemical. As described previously, toxicity data have historically been available for only a relatively small percentage of the 40,000 chemicals in commerce today. This lack of available toxicity data is, in part, due to historical regulations which did not require extensive testing prior to chemical use, but is also due to the fact that our understanding and knowledge of chemical testing and toxicity has evolved over time.

The subsections below explore the following points:

- Recent developments in toxicology and ecotoxicology including our understanding and analysis of existing data
- Regulatory forces that have driven the collection of data
- Special classes of chemicals defined under chemical control regulations that may emerge as contaminants.

3.2.1 DEVELOPMENTS IN TOXICOLOGY AND ECOTOXICOLOGY

As described in Chapter 2, most early efforts to characterize the toxicity and eco-toxicity of environmental pollutants within a regulatory context focused on just a few types of hazard. Early characterizations of environmental hazards focused on freshwater algae and cyanobacteria (growth inhibition), *Daphnia sp.* (acute immobilization), and fish (acute toxicity). Scientists evaluating human health concerns initially focused on the potential for acute toxicity in humans, carcinogenicity, and, to a lesser extent, certain forms of subchronic or chronic toxicity.

Over time, new test methods have been added which provide a more nuanced understanding of hazard. Newer test methods and risk assessments cover a more robust range of biotic system tests (e.g., vegetative, fish, and honeybee tests) and endpoints (e.g., endocrine disruption, sensitive populations). In addition, as noted in Section 2.4.2, the methods and understanding of toxicology and toxic endpoints have continued to evolve since US EPA first developed the IRIS database in 1985. Along with advances in laboratory testing methods, data analysis approaches have continue to progress with increased use of physiologically based pharmacokinetic (PBPK) and other meta-data models. Moreover, as public interest and outrage has continued to increase, regulatory agencies have often responded with approaches that rely on the most conservative interpretations of existing data.

The following examples demonstrate how existing toxicology data can be used by regulatory agencies to identify different outcomes regarding chemical toxicity and how those varying analyses can lead to vastly different regulatory requirements.

In April 2018, the California Department of Toxic Substance Control (DTSC) added a requirement to the California Code of Regulations that required any human health risk assessment to use Office of Environmental Health and Hazard Assessment (OEHHA) toxicity values if they were more conservative than current federal (i.e., US EPA IRIS) toxicity values. This had the effect of lowering cleanup levels for contaminated sites if State-specific values were lower than federal values. This regulatory change strongly affects the hazard assessment of tetrachloroethylene (PCE).

In 2016, OEHHA had updated the inhalation toxicity values for PCE using a lifetime rodent cancer study (JISHA 1993) and a recent PBPK model update (Chiu and Ginsberg 2011) [62]. Based on this assessment, OEHHA developed a unit risk factor (UR) for noncarcinogenic health effects upon inhalation of PCE of 6.1×10^{-6} per micrograms per cubic meter $(ug/m^3)^{-1}$ and a cancer slope factor (CSF) of 2.1×10^{-2} milligrams per kilogram body weight per day $(mg/kg\text{-}d)^{-1}$. Under the 2018 DTSC requirements, these values had to be taken into consideration in setting cleanup levels for sites.

US EPA initially entered PCE into IRIS in 1987 and updated the entry in 1987 and 1988.[63] Those values remained unchanged for two decades, until the Agency updated the toxicological assessment for PCE in 2012, using the same cancer study and PBPK model as OEHHA. However, even though the same study and model were used, US EPA developed a new UR of 3×10^{-7} $(ug/m^3)^{-1}$ and a CSF of 2×10^{-3} $(mg/kg\text{-}d)^{-1}$. These values are approximately an order of magnitude less conservative than the values derived by OEHHA from the same data [64].

Both US EPA and OEHHA agreed that the 2011 new PBPK model was an improvement over previous models; US EPA described the improvements as including

(1) the utilization of all the available toxicokinetic data for tetrachloroethylene and its metabolites in mice, rats, and humans; (2) the incorporation of available information on the internal toxicokinetics of TCA [trichloroacetic acid] derived from the most current PBPK modeling of trichloroethylene and TCA; and (3) the separate estimation of oxidative and conjugation metabolism pathways [GST pathway] [65].

Accurate modeling of these pathways was critical as PCE's carcinogenic potency likely depends upon the formation of genotoxic metabolic products.

Despite this agreement in study and analysis methods, US EPA and OEHHA still diverged in their final analysis and review of the data, specifically related to one of the metabolism pathways. Data suggested a much more uncertain outcome for the gluaththione-S-transferase-catalyzed conjugation pathway, especially in humans. Given the uncertainty in the data, when run for humans the gluaththione-S-transferase model predictions can differ by 3,000 fold, yielding significantly different results. Given this range, US EPA stated, "At this point, it is not possible to disentangle the contributions of uncertainty and variability to the very large range of estimates of tetrachloroethylene GSH [glutathione] conjugation in humans" [66]. US EPA therefore prioritized the dose metrics based on oxidative metabolism in the final analysis.

In contrast, OEHHA relied on the model outcome that indicated a much greater incidence of metabolism via the GSH pathway. Scientists from the DTSC acknowledged the very high variability in the GSH conjugation pathway results but concluded that the diverse ethnic population of California necessitated the use of the most conservative estimates regarding metabolism of PCE [61]. In other words, given the diverse ethnic make-up of California, it was necessary to use a model that accounted for the full impact of human inter-individual variability in genetic and other factors (i.e., diet, gender, age). As a result, OEHHA ran the PBPK model using both the oxidative and GSH conjugation metabolic pathways. This resulted in an order of magnitude more conservative toxicity values and eventually more conservative screening values. As a point of comparison, the US EPA residential indoor air screening value at a 1×10^{-6} cancer risk is $11ug/m^3$; DTSC uses the OEHHA toxicity criteria and developed a residential indoor air screening value at a 1×10^{-6} cancer risk of $0.46ug/m^3$.

Although PCE may not be considered an emerging contaminant per se, this example demonstrates how evolving differences in the understanding, evaluation, and application of toxicological data can lead to significantly different estimates of toxicity and risk. The potential for different interpretations of toxicological data is evident in many other chemicals. Two additional examples include 1,4-dioxane and trichloroethene (TCE).

US EPA released toxicological assessments for 1,4-dioxane in 2010 (oral exposure) and 2013 (inhalation exposure) and developed both noncancer and cancer oral and inhalation toxicity values based on these assessments [67]. At the time of the assessment, scientists debated the mode of action (MOA) of 1,4-dioxane. US EPA's standard and default approach is to assume that chemicals that cause cancer act through a nonthreshold or linear MOA. This default approach is typically very conservative as it assumes any exposure of the chemical could cause cancer, even at very low doses.

At the time of the assessments in 2010 and 2013, US EPA had information that indicated generally negative results for 1,4-dioxane in a number of genotoxicity assays. Toxicologists often interpret such results as evidence that the chemical does not cause cancer by damaging genetic material or DNA. Chemicals that are not genotoxic may still cause cancer, but it occurs through a different mechanism such as disruption of cellular metabolic processes. Despite this information, US EPA stated that [67]:

> The linear approach is recommended if the mode of action of carcinogenicity is not understood ... In the case of 1,4-dioxane, the mode of carcinogenic action for liver tumors is not conclusive. Therefore, a linear low-dose extrapolation approach was used to estimate human carcinogenic risk associated with 1,4-dioxane oral exposure.

The resulting cancer slope factor led to the calculation of a 1,4-dioxane risk-based drinking water limit of $0.3ug/L$ at a cancer risk of 1×10^{-6}.

As part of the toxicological assessment, US EPA did acknowledge that several members of the external peer review panel disagreed with this assessment and recommended the use of a nonlinear extrapolation approach to estimate human carcinogenic risk based on a threshold MOA. This recommendation was based on numerous nonpositive short-term in vitro and a few in vivo tests for 1,4-dioxane-induced genotoxicity, as well as an MOA hypothesis involving 1,4-dioxane-induced cell proliferation [67].

Since the publication of the 1,4-dioxane toxicological assessments, several other researchers have reviewed the toxicity data for 1,4-dioxane and determined that it does most likely cause liver tumors in laboratory animals through cytotoxicity followed by regenerative hyperplasia [68–70]. In other words, cancer occurs through a threshold or nonlinear MOA.

Many of the questions regarding the MOA for 1,4-dioxane were raised due to conflicting study results [69]. A closer review of the data from these studies demonstrated that the conflicting results were actually due to the study methods used at the time: pathologists generally just recorded the most severe pathology and other tumors were not recorded. As a result, important information on histological findings went unrecorded. Without this information, researchers and regulatory bodies could come to different conclusions about the chemical's MOA.

Was this new information sufficient to change the regulatory understanding of the MOA and toxicity of 1,4-dioxane? Using this same set of data, Health Canada developed a maximum acceptable concentration in drinking water of 50ug/L to be protective of public health [71] using a threshold or nonlinear MOA. This compares to the US EPA risk-based screening level of 0.3ug/L at a cancer risk of 1×10^{-6} based on the nonthreshold or linear MOA. As of this writing in 2019, US EPA is re-evaluating all the data for 1,4-dioxane and, as shown above, the interpretation and understanding of the MOA will be critical for driving regulatory requirements. In summary, applying current scientific understanding to data collected decades ago can result in new conclusions regarding hazard and risk; when those new conclusions change cleanup goals by an order of magnitude or more, they can either diminish previous concerns or cause new site concerns to emerge.

The final example of toxicological data and analysis driving emerging contaminants is TCE. Many readers may be familiar with the concern over short-term exposure to TCE by a pregnant woman, which US EPA based on the results of a single study [72] to indicate that exposure of a pregnant woman in her first trimester may cause heart malformations or cardiac heart defects (CHD) in the fetus. In response to the findings from this study, US EPA released a policy related to vapor intrusion and short-term exposures [73] that led to many additional and urgent vapor intrusion investigations at Federal and State hazardous waste sites across the country [74].

Since that time several studies have been published both in support of and against the use of the findings of the study showing fetal heart malformations or CHD [75,76]. Most recently, the DOD proposed an Occupational Exposure Limit of 350ug/m^3 for TCE to be protective of all exposures including those related to vapor intrusion [77]. In this analysis, DOD eliminated the study showing fetal heart malformations or CHD from the analysis due to poor study quality. Consequently, the DOD value is two orders of magnitude higher than the US EPA's noncancer urgent indoor air level for a commercial work place of 8ug/m^3. The risk assessment and development of occupational exposure limits for TCE under the European Union regulation Registration, Evaluation, Authorisation and Restriction of Chemicals (REACH) provides additional perspective. It too dismisses the study showing fetal heart malformations or CHD based on the study quality [78]. Again, this example illustrates how the interpretation and use of a single study led to a vastly different understanding of potential toxicity and changed how the chemical was eventually managed.

Through the Toxic Substances Control Act (TSCA) review of the risks from existing chemicals (see Section 3.2.2), US EPA will be reviewing the available toxicological information for the three chemicals described above. Specifically, TSCA requires the US EPA to evaluate the best available science and not just rely on the previous assessment completed under IRIS. For PCE, it is possible that US EPA could revise the PBPK modeling to use the approach recently taken by California. This would increase the concern over exposures to PCE in the environment. In contrast, the assessment of new data for 1,4-dioxane and TCE could lead to less conservative results and these chemicals may no longer be a primary target as an emerging contaminant.

3.2.2 REGULATORY FORCES

Next we will examine the regulatory forces in the European Union and United States which have driven new assessments of the risks of exposure to chemicals. Those assessments create the potential for "new" contaminants to emerge and possibly, as described above, for previous concerns to abate.

3.2.2.1 Developments in the European Union

REACH, which entered into force on June 1, 2007, revolutionized chemical regulation because of the requirement to generate data. Under this regulation, manufacturers and importers must characterize the physicochemical properties, environmental fate, toxicological effects, and ecotoxicological effects of a chemical before it can be placed on the market. In addition, the data must meet regulatory standards for reliability. The data required to characterize a chemical on the market in the EU depend upon the tonnage of that chemical in commerce (tonnes per annum, or tpa on a per legal entity, per site basis). Table 3.3 summarizes the data requirements [79].

As of June 2019 manufacturers and importers have amassed data for 22,221 unique substances [80]. Of these chemicals, and per the requirements indicated in Table 3.3, substantial data have been collected for substances in the higher tonnage bands:

- 2,458 substances are manufactured or imported at >1,000 tpa
- 2,205 substances are manufactured or imported at 100–1,000 tpa
- 2,767 substances are manufactured or imported at 10–100 tpa.

TABLE 3.3

Data Requirements under REACH Based on Annual Tonnage

Study	Annex VII 1–10 tpa	Annex VIII 10–100 tpa	Annex IX 100–1,000 tpa	Annex X > 1,000 tpa
Physicochemical Properties				
State of Substance	✓	✓	✓	✓
Melting/Freezing Point	✓	✓	✓	✓
Boiling Point	✓	✓	✓	✓
Relative Density	✓	✓	✓	✓
Vapor Pressure	✓	✓	✓	✓
Surface Tension	✓	✓	✓	✓
Water Solubility	✓	✓	✓	✓
Partition Coefficient n-Octanol/Water	✓	✓	✓	✓
Flash Point	✓	✓	✓	✓
Flammability	✓	✓	✓	✓
Explosive Properties	✓	✓	✓	✓
Self-Ignition Temperature	✓	✓	✓	✓
Oxidizing Properties	✓	✓	✓	✓
Granulometry	✓	✓	✓	✓
Stability in Organic Solvents and Identity of Relevant Degradation Products			✓	✓
Dissociation Constant			✓	✓
Viscosity			✓	✓
Toxicological Information				
Skin Irritation	✓	✓	✓	✓
Eye Irritation	✓	✓	✓	✓
Skin Sensitization	✓	✓	✓ (in vivo)	✓ (in vivo)
Mutagenicity	✓	✓	✓ (+*in vivo*)	✓ (+*in vivo*)
Acute Oral Toxicity	✓	✓	✓	✓
Acute Dermal Toxicity		✓	✓	✓
Acute Inhalation Toxicity		✓	✓	✓
Repeat Dose Oral Toxicity (28-day)		✓	✓	✓ (> 12 months if warranted)
Subchronic Toxicity (90-Day)			✓	✓
Reproductive/Developmental Screening		✓		
Reproductive/ 2-Geneneration Reproductive testing			✓/if needed	✓/if needed
Developmental			✓ (prenatal)	✓

(*Continued*)

TABLE 3.3 (Cont.)

Study	Annex VII 1–10 tpa	Annex VIII 10–100 tpa	Annex IX 100–1,000 tpa	Annex X > 1,000 tpa
Toxicokinetics		✓ (if available)	✓ (if available)	✓ (if available)
Carcinogenicity				✓
Environmental Fate and Ecotoxicological Information				
Short-term Invertebrate (*Daphnia*) Toxicity	✓	✓	✓	✓
Algal Growth Inhibition	✓	✓	✓	✓
Short-term Fish Toxicity		✓	✓	✓
Activated Sludge Respiration Inhibition		✓	✓	✓
Daphnia Long-term Toxicity			✓	✓
Long-term Toxicity Testing on Fish			✓	✓
Bioaccumulation in Fish			✓	✓
Degradation (Biotic)	✓	✓	✓	✓
Abiotic Degradation		✓	✓	✓
Hydrolysis as Function of pH		✓	✓	✓
Identification of Degradation Products			✓	✓
Soil Adsorption/Desorption Screening Test			✓	✓
Effects on Terrestrial Organisms: Short-term toxicity to invertebrates; effects on soil organisms; short-term toxicity to plants			✓	✓
Long-term Effects on Terrestrial Organisms: invertebrates, plants, soil organisms, reproductive toxicity to birds				✓

The registration data for these compounds, summaries of which are publicly available, provide new insights into the toxicity and ecotoxicity of many chemical compounds. Before REACH, many of this data had not been collected or simply were not publicly available.

From this knowledge, new concerns about environmental contamination may emerge. Recall, for example, that the IRIS database contains information on just 568 chemicals, and that US EPA developed most of the associated dose–response data before 1995. These new data obtained under REACH may well change the previous understanding of the hazards of exposure to chemicals in the environment. And regulatory developments in the United States, as described below, provide the catalyst for re-examining long-held understandings of contaminant toxicity.

3.2.2.2 Developments in the United States

As previously discussed, limited amounts of data were available as a result of the 1976 TSCA. Although TSCA provided some mechanisms for US EPA to collect data and regulate chemicals, US EPA could only require toxicity testing if an unreasonable risk were likely.

The Frank R. Lautenberg Chemical Safety for the 21st Century Act (LCSA), which amended TSCA in 2016, changed the historical approach for regulating chemicals. (TSCA as amended by LCSA is simply referred to as TSCA throughout the remainder of this section.) In brief, since 2016 TSCA requires the US EPA to assess the risks from previously unassessed chemicals using the best available data. Further, new TSCA requirements sharpen the criteria for evaluating risk.

As a first step, the amended law directs US EPA to compile and publish a list of the chemicals in commerce in the United States. In February 2019, US EPA determined that 40,655 chemicals were actively used in, produced in, or imported into the US in the 10 years prior to June 2016. Few of these chemicals have been regulated to date, either under TSCA or under other environmental laws. Of the approximately 62,000 chemicals on the market when TSCA originally came into effect in the 1970s, the US EPA regulated fewer than a dozen chemicals or chemical categories to control risks during manufacture and use; of the approximately 22,000 chemicals brought to market since 1976, the US EPA has only regulated the manufacture and use of 2,844 [81] as of June 2019. Many of the chemicals on the market simply have not been fully assessed for hazard or risk.

Only a very small fraction of the chemicals in commerce are regulated or even regularly sampled for in drinking water or as part of evaluation of media at sites undergoing testing and cleanup (See Chapter 2). For example,

- The Occupational Safety and Health Administration (OSHA), whose mission is "to assure safe and healthful working conditions for working men and women by setting and enforcing standards" [82], has only developed 470 Permissible Exposure Limits (PELs) for approximately 300 chemicals or 0.75% of the 40,000 active chemicals [83].
- The US EPA IRIS database currently contains toxicity data and values for 568 chemicals, also a small percentage (1.4%) of the chemicals currently listed on US EPA Active TSCA inventory.(Section 2 contains additional information on the numbers of chemicals regulated under various environmental laws in the United States.)

To address potential gaps in chemical regulation, TSCA (as amended in 2016 with LCSA) outlines a detailed process with timelines for evaluating the thousands of chemicals on the TSCA Inventory. In short, the Agency must prioritize chemicals for evaluation and then methodically complete those evaluations on a schedule set by Congress in the amendments to TSCA. Figure 3.6 illustrates this timeline.

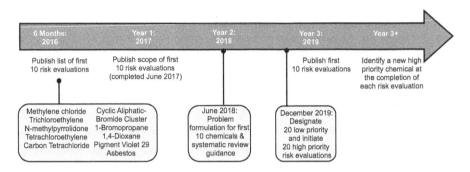

FIGURE 3.6 Chemical Evaluation Process by US EPA under the LCSA Amendments to TSCA

The US EPA's starting point for prioritizing chemicals is the *TSCA Work Plan for Chemical Assessments: 2014 Update* (2014 Work Plan) [84]. The 2014 Work Plan identified 90 chemicals that exhibited one or more of the following:

- Potential concern for children's health
- Neurotoxic effects
- PBT
- Probable or known carcinogen
- Used in children's products
- Detected in biomonitoring programs.

US EPA then ranked each of the identified chemicals in the 2014 Work Plan based on toxicity, exposure, and bioaccumulative potential using a 1 to 3 scale with 3 being the most toxic or bioaccumulative or having the highest exposure. The maximum possible score, considering all 3 factors, is 9.

In September 2018, US EPA released *A Working Approach for Identifying Potential Candidate Chemicals for Prioritization*. This document describes the approach US EPA will take including the data to be considered to support the prioritization process. In particular, US EPA clarified that the "working approach is to primarily look to the 2014 Work Plan for high-priority potential candidates as TSCA requires that at least 50% of the chemicals undergoing risk evaluation as of December 2019 must come from the 2014 Work Plan" [85]. The prioritization process outlined by US EPA indicates that in addition to the three variables outlined in the 2014 Work Plan (i.e., toxicity, exposure, and bioaccumulative potential), US EPA will also consider the priorities of US EPA and other federal agencies, the quality and quantity of available data, US EPA's workload, and US EPA's ability to meet statutory deadlines.

Once a chemical is identified as high priority, US EPA has up to three and a half years to complete a risk evaluation. If the risk evaluation identifies a potential injury to human health or the environment (excluding consideration of cost and considering susceptible subpopulations), US EPA must take actions to prohibit or limit the chemical use. Once a risk evaluation for a high priority chemical is complete, TSCA requires US EPA to add another high priority chemical. In other words, the evaluation of high priority chemicals will be an ongoing process for many years to come.

As illustrated in Figure 3.6, in December 2016 US EPA identified the first 10 high priority chemicals for evaluation based on the Work Plan. US EPA published the next 20 high and low priority chemicals for evaluation under TSCA on March 21, 2019 [86]. All 20 high priority chemicals were from the 2014 Work Plan and scored either a 6, 7, or 8 in the 2014 Work Plan. Although 23 chemicals that scored an 8 or 9 in the Work Plan were still available for evaluation, US EPA did not solely consider the Work Plan score. Instead, US EPA also evaluated data availability and chemical groupings as a way to streamline the risk evaluation process. For example, on the list of 20 high priority chemicals are 5 chemicals classified as a phthalate ester, 3 halogenated flame retardants, and 7 chlorinated solvents. Eventually using a prioritization and binning approach, US EPA will identify other high priority chemicals for evaluation [85].

US EPA has provided information on the overall approach it will undertake to identify additional high priority chemicals; however, the Agency has not yet released detailed information as of June 2019. There are, though, many other lists to consider where US EPA or other regulatory bodies have identified chemicals that are tracked due to the quantities currently produced or used or due to information that indicates the chemical may have some human or ecological toxicity.

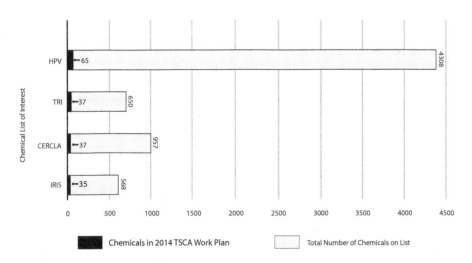

FIGURE 3.7 Comparison of 2014 TSCA Work Plan Chemicals to other Chemical Lists of Interest

Figure 3.7 shows the results of several comparisons of the chemicals in US EPA's Work Plan to other lists discussed in this book. The comparisons shown in Figure 3.7 lead to the following findings:

- The majority (>50%) of the chemicals on the 2014 Work Plan are HPV chemicals, indicating that they are produced or imported at greater than 1 million pounds per year. This comparison indicates that while many of the Work Plan chemicals are produced in large quantities, there are many other HPV chemicals that are not identified by US EPA as a priority for evaluation.
- Only about one-third of the chemicals in the 2014 Work Plan have any toxicological data in IRIS, regardless of the year of IRIS entry. The TSCA risk evaluation requires the use of the best available science and for the majority of chemicals this is likely to result in a potentially significant change in the understanding of estimated toxicity. Two of the first 10 chemicals evaluated under TSCA had no information in IRIS, 3 had entries that were more than 10 years old, and 5 had entries that were less than 10 years old. Looking at the rest of the 2014 Work Plan chemicals identified for evaluation [86], 45% have no entries in IRIS, 52% have entries in IRIS that are more than 10 years old, and only 3% have entries that are less than 10 years old.
- Approximately 40% of the 2014 Work Plan chemicals are also tracked and reported on the TRI, indicating that for at least some of the Work Plan chemicals, US EPA has use, release, and emissions data by which a risk evaluation could be completed.
- Thirty-seven of the CERCLA list chemicals are also on the 2014 Work Plan. This comparison demonstrates that approximately two-thirds of the chemicals on the 2014 Work Plan are not currently sampled for at Superfund sites. If the risk evaluation under TSCA identifies previously unrecognized hazards, then "new" contaminants not currently part of site investigations may emerge.

Overall, this information clarifies that while US EPA currently has a clear process in place to evaluate chemicals under TSCA, there is no standardization in other regulatory lists or a single "emerging contaminant" list that can be identified and checked. There are many different drivers to each of the regulatory lists discussed in this book and they each may provide some information that can be used towards understanding emerging contaminants.

3.2.3 Special Cases: PBT and PMT/PMOC Compounds

Compounds that can persist in the environment, particularly if they are toxic, present special concern. Two distinct classes of compounds are relevant to the topic of emerging contaminants and are under scrutiny by regulatory agencies: PBT compounds and persistant, mobile and toxic compounds (sometimes

TABLE 3.4

Criteria for PBT Compounds

Jurisdiction	Persistent	Bioaccumulative	Toxic
United States [87,88]	$t_{1/2, \text{water, soil, or sediment}}$ $\geq 2\text{–}6$ mos	BCF >5,000 (1,000–5,000 moderate potential)	Human health: systemic toxicity, mutagenic damage, reproductive toxicity, or developmental toxicity; Aquatic toxicity: • low concern: fish chronic value (ChV)* >10mg/L • moderate concern: fish ChV 0.1–10mg/L • high concern: fish ChV <0.1mg/L
European Union [89]	Fresh water: $t_{1/2, \text{water}}$ >40d $t_{1/2, \text{sediment}}$ >120d; Marine: $t_{1/2, \text{water}}$ >60d $t_{1/2, \text{sediment}}$ >120d	BCF 2,000L/kg, or Log k_{ow} >4.5	Human health: substance is classified as carcinogenic (category 1A or 1B), germ cell mutagenic (category 1A or 1B), or toxic for reproduction (category 1A, 1B, or 2), or there is other evidence of chronic toxicity, as evidenced by classifications for specific target organ toxicity after repeated exposure according to Regulation EC No 1272/2008; Aquatic toxicity: NOEC or EC10 <0.01mg/L for marine or freshwater organisms
Stockholm Convention [90]	$t_{1/2, \text{water}}$ >2mos $t_{1/2, \text{soil}}$ >6mos $t_{1/2, \text{sediment}}$ >6mos Long-range transport if $t_{1/2, \text{air}}$ >2d	BCF or BAF (aq) >5,000, or Log k_{ow} >5	"(i) Evidence of adverse effects to human health or to the environment that justifies consideration of the chemical within the scope of this Convention; or (ii) Toxicity or ecotoxicity data that indicate the potential for damage to human health or to the environment."

* US EPA defines ChV as follows [91]: "The ChV, or Chronic Value, is defined as the geometric mean of the no observed effect concentration (NOEC) and the lowest observed effect concentration (LOEC)."

designated as persistant mobile organic compounds). Each category is discussed below.

3.2.3.1 PBT compounds

Persistent, bioaccumulative and toxic compounds are, as the name implies, difficult to attenuate naturally, may be difficult to remediate, and are likely to
have relatively low cleanup goals as a result of their toxicity. Readers will
intuitively grasp why such compounds are of concern based on experience
with polychlorinated biphenyls (PCBs), arguably the best known of PBT compounds. But most readers probably do not appreciate the number of chemicals
in this category that could potentially emerge as environmental contaminants.

3.2.3.2 What Makes a Chemical PBT?[4]

Table 3.4 shows the criteria for defining a chemical substance as PBT according to regulators in the United States and European Union; it also contains
related criteria under the Stockholm Convention, an international treaty
regarding Persistent Organic Pollutants. Authorities differ on the appropriate
benchmarks for defining a chemical compound as PBT, as illustrated in Table
3.4. (And in fact, some scientists believe that PBT criteria should be updated
to reflect advances in scientific understanding [92].) We'll examine the factors
defining persistence, bioaccumulation, and toxicity below (where BCF is Bioconcentration Factor and BAF is Bioaccumulation Factor).

Persistent compounds do not readily degrade in the environment by biodegradation, hydrolysis, or photolysis. Of these three processes, biodegradation is
generally the most significant. An organic substance resists biodegradation in
two general cases: if it is hydrophobic or if the chemical bonds in the substance cannot be readily broken. The first case pertains to hydrophobic substances that are poorly soluble in water and tend to sorb strongly to soil or
sediment. Such characteristics limit the availability of a compound for micro-
bial degradation [93]. In the second case, the bonds between the atoms in
a molecule are simply too strong to break through unenhanced biologically
mediated reactions. (In some cases, enhancing biodegradation within an engin-
eered treatment system can effect degradation.)

Degradation is often modeled as a pseudo first-order reaction

$$C = C_0 e^{-kt}$$

Where t = time, C = concentration at specified time, and k = rate constant.

Then the rate of biodegradation, or conversely, a chemical's persistence in
the environment, can be characterized by the amount of time it takes to
reduce the concentration by half. The "half-life" ($t_{1/2}$) for a chemical differs
on whether it is found in air, water, soil, or sediment, and also depends upon
temperature.

PBT chemicals are considered to be persistent because they have long half-
lives in the environment. Although different authorities define persistence differ-
ently, in general a chemical with a half-life in the order of a few months or
more is considered to be persistent. The half-life of PCBs, for example, depends

upon the degree of chlorination and can range from 2 to 330 months in water and 36 to 460 months in sediment at an average annual temperature of 7°C (approximately 46°F) [94]. Table 3.4 contains one notable difference with respect to characterizing PBT chemicals by their half-life. In contrast to the framework offered by the Stockholm Convention, regulators in the US and EU do not consider the potential for long-range transport by air, as represented by a half-life in air, when defining PBT compounds. Thus the corresponding regulations on PBT chemicals do not necessarily address the compounds that may migrate through air to the Arctic, as discussed in Section 3.1.3.

Bioaccumulation refers to the net uptake of a substance from the environment by respiration, ingestion, and dermal absorption from any source, including the ingestion of other organisms [95]. Alternatively, some scientists measure bioconcentration, the uptake of a chemical substance by all routes other than ingestion [96,97]. These two phenomena are characterized by the BCF and BAF, respectively, which represent the ratios between the concentration of the substance in the organism and in the environment. Hydrophobic (or lipophilic) compounds characterized by a relatively high value of the octanol-water partition coefficient k_{ow} are more likely than other chemicals to bioaccumulate. As shown in Table 3.4, authorities have set criteria for determining whether a chemical is bioaccumulative based on its BCF, BAF, and/or the logarithm of the octanol-water partition coefficient (log k_{ow}). To illustrate this characteristic for PCBs, the well-known PBT family of compounds, consider that one literature review found that the values of log BCF for PCBs ranged from 3.11 to 5.97, depending upon the degree of chlorination; similarly, values of log K_{ow} ranged from 4.63 to 8.18 [98].

A chemical is considered to be toxic to humans, within this context, if it is a carcinogen, mutagen, or reproductive toxicant. Some authorities use additional criteria. As noted in Table 3.4, in the European Union a chemical can be classified as "'T' if [t]here is other evidence of chronic toxicity, as identified by the classifications: T [toxic], R48 [Danger of serious damage to health by prolonged exposure], or Xn, R48 according to Directive 67/548/EEC". Also as shown in Table 3.4, toxicity may also be gauged based on hazards to aquatic life.

3.2.3.3 How Many PBT Compounds Do We Need to Consider?
The number of PBT compounds that are or have been in commerce is unclear. Strempel et al. [99]. attempted to estimate the universe of potential PBT compounds. They screened approximately 95,000 chemicals that have been or are now on the market in the European Union by comparing measured or estimated properties to the criteria used to classify PBT chemicals in the European Union. They found that over 2,900 chemicals could potentially be considered to be PBT. The most common chemical structures in this group included chlorinated and brominated aromatic systems; chlorinated and brominated cycloaliphatic compounds; highly branched alkyl substances and aromatic compounds with several highly branched alkyl, ether, or tertiary amine groups as substituents; triphenylmethyl substances; spiro compounds; various per- and polyfluorinated alkyl substances of different chain lengths; compounds with trifluoromethyl substituents; nitroaromatic compounds; tertiary amines with highly branched alkyl

groups; and polycyclic aromatic hydrocarbons. Fifty-seven of the potential PBT compounds were considered to be HPV chemicals in the European Union.

The list of potential PBT compounds identified by Strempel et al. was compared to two lists relevant to the United States: chemicals identified as HPV, i.e., manufactured at more than 1 million pounds per year, and chemicals listed as hazardous substances under CERCLA. Fifty chemical compounds are manufactured or imported into the United States at >1 million pounds per year and that could be considered PBT compounds; four of these substances are on the CERCLA list, and five of these chemicals are tracked by TRI. Of these compounds, only three PBT chemicals manufactured or imported into the United States at >1 million pounds per year are tracked by TRI and are listed as a hazardous substance under CERCLA.

This comparison is cause for further thought but not alarm. Contaminants will emerge from this cohort of 50 compounds (or from the universe of PBT compounds in commerce at less than HPV levels) only if they have been released to the environment in substantial concentration. Many may be or have been used under controlled conditions, e.g., as intermediates in manufacturing, and therefore releases to the environment may have been limited.

Pentachlorothiophenol (abbreviated PCTP; CAS No. 133-49-3) illustrates the case of a HPV/PBT substance used in an intermediate step in manufacturing. According to information compiled by the US EPA [100,101], PCTP has been used as a mercaptan cross-linking agent to make rubber more pliable, notably to make golf balls. The Agency has speculated that releases may have occurred during manufacturing to water (from process wastewater or equipment cleaning), and perhaps from the disposal of off-spec product; releases to air were unlikely due to the low vapor pressure of PCTP.

The US EPA has not been able to identify data to quantify releases to the environment. PCTP is not a TRI chemical; the US EPA has acknowledged that it has no data on the disposal of PCTP and has not identified any studies of PCTP in the environment. The US EPA identified two biomonitoring studies which found PCTP in human urine samples, but the scientists who performed both of those studies acknowledged that PCTP may have been present as a result of the biotransformation of hexachlorobenzene (another PBT compound) stored in tissues [102,103].

One would expect that the use of PCTP in the manufacture of rubber – rather than in some use which might result in dispersion thoughout the environment – would limit the potential for PCTP to emerge as an environmental contaminant of concern. But the US EPA identified no data to test that hypothesis.

The global use of PCTP declined with the realization that it degrades to form several teratogenic compounds. As a result, many countries ban PCTP. The potential for releases of this chemical in the present day are limited. While PCTP has been an HPV substance in the United States, it is no longer. In fact no company reported manufacture and/or import of PCTP above 25,000 pounds (approximately 11,300 kilograms) per site in the 2016 Chemical Data Reporting to US EPA [104].

This case study illustrates limitations on our ability to anticipate with certainty developments in the realm of emerging contaminants. The screening comparison of HPV chemicals to PBT chemicals identified PCTP as worthy of attention. Its use in chemical processing, rather than in a dispersive sense, leads to the hope that – considering its degradates are teratogenic – it is not a signficant environmental contaminant. But without data, one cannot confirm that conclusion. And it is not always practical to collect representative data to address every hypothesis about every potential contaminant.

3.2.3.4   Persistent, mobile and toxic compounds

Scientists have recently called attention to the class of compounds considered to be persistent, mobile and toxic, abbreviated PMT. Some researchers refer to persistent, mobile organic compounds (PMOC). While analogous to PBT chemicals in their persistence and toxicity, they differ in one important feature: these molecules, largely polar organic compounds, tend to be soluble, migrate readily through ground water and surface water, and pass through conventional wastewater treatment plants [105].

The solubility or mobility thresholds for characterizing a chemical as PMT or PMOC have not been unequivocally defined. The German Environment Agency proposed in 2017 that persistence and toxicity be characterized for this purpose as described for the European Union in Table 3.4 and that the following criteria [106] should be considered based on solubility in milligrams per liter, carbon-water partition coefficient (K_{oc}), and the pH-dependent octanol-water partition coefficient (D_{ow}):

- the highest water solubility is ≥ 0.15mg/L and the lowest log Koc is ≤ 4.0 at environmentally relevant pH range of 4–9 and at a temperature of 12°C
- in the absence of log K_{oc} data, the highest water solubility is ≥ 0.15mg/L and the lowest log D_{ow} is ≤ 4.0 at environmentally relevant pH range of 4–9 and at a temperature of 12°C (54 °F).Alarms have begun to sound in the European Union about the threat that such chemicals pose to drinking water supplies [107,108]. In short, this loosely defined class of contaminants is beginning to emerge based upon concerns over exposure through drinking water. While some such as MTBE and PFAS are recognized contaminants of soil and groundwater, others are not.

Teams of scientists in the EU have evaluated the chemicals registered under REACH to determine how many of the chemicals in commerce might be PMT/PMOC.

- Arp et al. (2017) [109] used an early version of the PMT criteria proposed by the German Environment Agency to evaluate organic chemicals registered under REACH circa 2014. They determined that 2,167 unique substances (including organic and pseudo-organic chemical compounds)

might be of concern: 1,811 PMOC compounds and 356 PMOC precursors with the potential to be hydrolyzed to PMOC.
- Schulze et al. (2018) [110] built on the work by Arp et al. by modeling
  the emission potential of 2,167 PMOC based on the uses and volumes in
  commerce reflected in REACH registrations. They estimated that 936
  PMOC are emitted into the environment in the EU and rank-ordered
  those chemicals.
- A team from the Norwegian Geotechnical Institute (2018) [111] assessed
  the 15,469 substances registered under REACH as of May 2017 using the
  PMT criteria proposed by the German Environment Agency with slight
  modification to the criteria used to define mobility. (They removed the
  water solubility criterion of 0.15mg/L and the temperature requirement of
  12°C based upon their assessment of the relevance of those factors.)
  Their preliminary assessment of substances registered under REACH
  circa 2017 that could be considered PMT substances identified 240
  compounds.

Each team noted limitations to their methods, data, and findings. Nonetheless, their preliminary work provides an initial view of contaminants that may
emerge from within this class. Routine analytical methods do not detect many
PMOC in water [112]. As analytical methods evolve, more PMOC may
emerge as contaminants of concern.

Of the 2,167 PMOC or precursors identified by Arp et. al., approximately
260 are manufactured in or imported into the United States at 1 million
pounds per year or more. These include compounds like 1,4-dioxane, which
has emerged as a contaminant of concern. Some, like carbon tetrachloride
and naphthalene, have been recognized and remediated at hazardous waste
sites for many years. However, 228 PMOC manufactured or imported in the
US at 1 million pounds per year or more are not on the CERCLA list of hazardous substances.

3.2.4 IMPLICATIONS FOR EMERGING CONTAMINANTS

Over the last 50 years, scientists' understanding of toxicology and regulators'
reflection of those findings have continued to evolve. New test methods and
nuanced means of interpreting data have changed scientists' ability to recognize potential hazards to human health or to creatures in the environment.
Over time, those changes become reflected in regulatory requirements and
regulatory submittals. With the recognition of previously unrealized hazards
comes the potential for new contaminants to emerge.

Early environmental hazard characterization often focused on freshwater
algae and cyanobacteria (growth inhibition), *Daphnia sp.* (acute immobilization), and fish (acute toxicity). Over time scientists developed test methods to
look at different species and more closely evaluate specific endpoints associated with toxicity, such as endocrine disruption and developmental effects.
Eventually, this led to newer analysis methods that could handle the many

studies and data points often available for individual chemicals. Meta analyses and PBPK modeling were developed to allow for a better understand and translation of laboratory animal studies into effects in humans including sensitive subpopulations (i.e., children and the elderly). Chemicals characterized as PBT (estimated at as many as 2,800) or PMT/PMOC and their precursors (in the order of 2,200 chemicals in commerce) are under increasing scrutiny.

Even data using the early test methods were often not available to regulators or to the public. Changes in regulations in the US and EU since 2006 are changing the availability of data and driving the assessment of risks. Under REACH, chemical manufacturers and importers have summited dossiers containing data for over 22,000 substances. Dossiers for approximately 2,800 of those chemicals provide foundational information on physical/chemical properties, environmental fate, and toxicity/ecotoxicity. The dossiers for nearly 4,700 substances provide richer data sets that characterize additional forms of toxicity and the effects on additional species in the ecosystem. All in all, the data compiled under REACH represent an unprecedented trove of data that provides new understanding of the potential hazards of exposure to many chemicals.

In the United States, 2016 amendments to TSCA require the US EPA to assess the potential risks from substances in commerce that are designated as a priority. The Agency must do so using the best available scientific data, which is remarkable given the generation and compilation of data under REACH. (Access to the details of those data may be limited by data ownership under REACH, but the summary information is freely available.) These assessments may revolutionize the US EPA's characterization of hazard. Recall that IRIS, the database of dose–response factors used to derive cleanup goals, originated in the 1980s. Most of the current entries reflect science of the 1980s and 1990s. Many focus on limited toxicological effects and lack supporting scientific information. As the US EPA assesses toxicity and ecotoxicity under TSCA, it may identify new contaminants of concern or the Agency may reassess "safe" levels and cleanup goals can change as a result.

New developments such as these in the understanding of exposure and hazard may cause a contaminant to emerge as a concern. But as the case studies in this book show, sometimes a contaminant emerges only after a long period of time and sometimes it does not emerge at all. To address that puzzle we'll look at one more factor: public outrage.

3.3 THE "X FACTOR": OUTRAGE

Outrage over chemical contamination is a powerful force. Consider the history of Alar. In 1989, the Natural Resources Defense Council released a report on Alar (trade name for daminozide), a growth regulator chemical sprayed on apple trees to prevent apples from falling before they ripen. The report estimated that Alar could cause cancer, putting children at risk. The television program *60 Minutes* broadcast the story and the ensuing public panic resulted in apples being removed from schools and the demand for organic apples spiking. In consequence conventional apple growers reported $1 billion in losses [113].

However, following this outcry, the toxicity assessment behind the initial study was found to be overly conservative, and several groups have indicated the study used questionable scientific evidence [114]. This incident quickly abated and the apple industry rebounded, but there were a number of lessons learned from this event. Among the reported lessons were these [113]:

- The public can quickly become alarmed by worse-case human health risk assessments.
- Too few of the media reporters understood the science behind the study and therefore inadvertently presented incorrect information.
- The cancer assay protocols may not evaluate a chemical's cancer potential in an absolute sense.
- Lack of transparency fuels public outrage – in the words of the author of a 1990 paper on Alar [113], "the chemical manufacturers should stop being defensive and secretive about their products as it promotes public distrust."

These lessons remain valid and, it can be argued, are even more applicable in this day of social media when public alarm can spread quickly.

A second example illustrates another important fact: as outrage ignites, it becomes more difficult to assess the balance between risks and benefits. Ethylene oxide (EO) is a regulated chemical that has been identified as an emerging contaminant based on permitted discharges. This gas emerged as the sterilization method of choice for medical devices due to its advantages over other methods such as the use of steam or gamma and electron beam irradiation [115]. EO is a flammable gas and has adverse human health effects. EO, by default, inherently must have high toxicity for it to be such an effective sterilization agent against bacteria, fungi/spores, and viruses. Data also indicate that EO is a potential carcinogen [116]. Worker exposure concerns were addressed through OSHA EO regulations passed in 1984, and subsequently the use of this sterilization agent became more prevalent, especially in modern plants where highly automated, controlled systems were installed.

In addition to the risks described above, sterilization using EO has benefits. EO has a high diffusivity and therefore can effectively sterilize medical devices within their packaging. EO is an effective bactericide, sporicide, and virucide agent. The use of EO has allowed for the development and use of sensitive single-use medical devices required for specific surgical procedures. EO grew to become the most cost-effective, low-temperature, and reliable sterilization process.[115]

In 2015, United States national aggregate production was estimated to be between 5 billion and 10 billion pounds (2.3 billion to 4.5 billion kilograms) per year [117]. In 2016, over 118 facilities performed TRI reporting on EO [118]. Those data indicated over 300,000 pounds (approximately 136,000 kilograms) of EO were legally released into the air via permitted discharge across several facilities, with over 17 million pounds of waste either treated, recycled, or burned for energy recovery, and over 300,000 pounds (approximately 136,000 kilograms) legally discharged on- or off-site.

News media began reporting on EO in 2018. They questioned whether EO emissions were responsible for the high cancer risk near a medical sterilization facility owned by Sterigenics located in Willowbrook, IL [119]. The public backlash and political pressure on the Illinois EPA became so intense that regulators temporarily shut the facility down in February 2019 [120], even after the company installed new air emission controls. At the time of shut-down, Sterigenics released a statement that said [121]:

> The Illinois EPA's actions to suspend operations at the Sterigenics Willowbrook facility are indefensible. Sterigenics Willowbrook has consistently complied with all state permits and regulations and Sterigenics has been in ongoing cooperation with the Illinois EPA and other officials regarding the safe operation of the facility.

> Unilaterally preventing a business that is operating in compliance with all state permits and regulations from carrying out its vital function sets a dangerous precedent. The Illinois EPA's decision will place the health and lives of thousands of patients who rely on the critical medical products sterilized at Willowbrook at risk.

In March 2019, the US Food and Drug Administration (US FDA) warned of the consequences of this shut-down: the risk of potential medical product shortages due to the closure of this large sterilization facility [122]. In its statement, the US FDA emphasized how sterilization of medical devices is critical to our healthcare system, and how EO is a commonly used method of sterilization that is considered safe and effective. They also recognized the environmental considerations that were affecting the manufacturer's ability to use this process and began working to secure alternative locations and methods for sterilization of devices to mitigate potential product supply issues. In this statement, however, they also indicated that Viant, another sterilization contractor, planned to close their Grand Rapids EO sterilization facility in 2019 after receiving attention on air quality issues from the Michigan Department of Environmental Quality. In April 2019, the US FDA released another statement that, despite attempts to prevent medical device shortages due to the plant shut-down, there was a temporary shortage of a specific type of tracheostomy tube [123] most likely to impact pediatric patients. The rapid timeframe to shortage is alarming as it can take up to six months to obtain regulatory approval to change sterilization methods, and many manufacturers have only one or two months of inventory on hand at any time [124]. As of this writing in June 2019, the Sterigenics facility remained closed, and litigation between Sterigenics and Illinois officials was ongoing.

Social scientists have examined the factors that cause outrage to ignite in cases like this one. While risk assessors consider risk to be a function of exposure and hazard, Peter Sandman famously coined the formula Risk =

TABLE 3.5

Components of Outrage that Influence Public Perception

Risk Component	Example
Voluntary vs Coerced	Choosing to be a smoker (and accepting the risks) versus being subjected to second hand smoke from nearby smokers (i.e., did not voluntarily accept risks)
Natural versus Artificial/Industrial	Ground apricot seed versus plastic microbeads in emolliating facial cleansers
Familiar versus Foreign/Exotic	Alcohol in beverages versus Bisphenol A (BPA) in water bottles
Memorable versus Forgettable	Rusty drums marked with toxic symbols stacked haphazardly outside a warehouse versus underground storage tanks
Chronic versus Catastrophic	Low-density, long-term plastic pollution versus visible large tracts of plastic garbage floating in the ocean
Known versus Unknown	Gasoline, in common use and a common contaminant at retail sites, versus PFAS becoming a broad contaminant (unknown characterictics, unknown toxicity data, unfamiliar to us all)
Personal Control	Driving a car versus being a passenger on a plane
Fairness	Risk spread evenly (gasoline retail sites) versus unevenly (neighborhoods located near factories with significant air emissions, e.g., outrage over EO)
Trust	News media versus social media, depending upon one's perspective.

Hazard + Outrage to explain public perceptions of risk [125]. He distinguished between two crucial concepts:

- Dangerous risk – the technical component which represents a hazard that can cause harm
- Upsetting risk – the social or cultural component which represents outrage. Dr. Sandman also estimated there is only an approximate 20% correlation between dangerous risk and upsetting risk [126].

Perception influences outrage. How we perceive risks will affect our reaction. Dr. Sandman identified the components which strongly influence how people perceive risk and generate outrage [125], the most relevant of which are shown in Table 3.5 with examples offered by the authors.

The pesticide dichloro-diphenyl-trichloromethylmethane (DDT) provides a powerful example of the effect of outrage. The public initially looked upon DDT with favor; when that opinion changed and outrage grew, it led to a ban on DDT and changed the way chemicals are regulated in the United States.

This story begins in 1948, when Paul Müller won the Nobel Prize in 1948 for discovering the use of DDT as an insecticide [127]. In his Nobel Prize award speech, Dr Müller described the successful role of DDT in quickly overcoming a typhus epidemic in 1944 [128]. But even as society recognized the benefits of DDT in stopping disease, the American Medical Association (AMA) began to raise alarms about the effect of pesticides and to suggest legislative efforts may be required to control the use of pesticides in order to protect the public [129].

Few heeded the warnings of the AMA, and DDT production grew by almost 20 times between 1948 and 1962 [130,131]. Then in 1962 Rachel Carson published the book *Silent Spring* [132] and a series of articles in *The New Yorker* based on the book. Written in language that translated scientific concepts into an understandable message, *Silent Spring* "made a powerful case for the idea that if humankind poisoned nature, nature would in turn poison humankind" [133]. These articles were widely read by the public, including President John F. Kennedy [134].

Silent Spring connected the public to the scientific discovery that DDT weakened bird eggshells, or in simple terms, that humans' use of chemicals damaged the environment [132]. This environmental message spread quickly. It catalyzed an environmental movement within the United States that ultimately led to the formation of the United States EPA in 1970 [133].

Manufacture and use of DDT peaked in the early 1960s, right around the publication of *Silent Spring*. The US EPA estimated that over 1 billion pounds of DDT were applied in agricultural and commercial use over the lifetime of DDT use [135]. As a result of the outcry over its hazards, restrictions on DDT began in 1969, and the US EPA completely banned DDT from use in the United States in 1972 [135].

Outrage has real impact, as illustrated by these case studies. With the rise of social media, outrage can spread rapidly. We've come a long way in technology since the days of *Silent Spring*, where magazines, books, newspapers, radio, and television were the main sources of information. Today, social media can disseminate information out to a broad, global audience at light speed. An alarming truth about the internet and social media is that the spread of misinformation is faster, deeper, and broader than factual and true information. In one notable study [136], social scientists evaluated the diffusion of true and false stories distributed on Twitter over an 11-year time period. They analyzed over 126,000 stories tweeted by approximately 3 million people more than 4.5 million times. The researchers found that the false news was more novel than the true news, suggesting that people are more likely to share false news. They also found that while robots accelerated the spread of news, they did this equally for false and true news. Together this indicates that false news spreads more rapidly and broadly than true stories because of people, not robots. The implications for emerging contaminants are ominous: if there is not enough information to quickly alleviate public concerns about a particular chemical or class of chemicals, misinformation may begin to spread and rapidly escalate public outrage. As outrage spreads and escalates it can lead legislators to take action on an emerging contaminant. What then makes an idea catch on? Social scientists have identified factors that make an idea contagious. Developed from observing the momentum that builds behind anything from the buzz about a new restaurant to trends in baby names, these principles also help to explain how societal concerns over emerging contaminants spread. These factors are [137]:

- Social currency: When people talk about a hot trend and seem "in the know", they can influence how others see them and can gain social status. This does not mean that bloggers writing about contamination issues do so to gain attention, but nonetheless talking about the latest emerging contaminant can reinforce visibility within a community and reflect positively on those who spread the word.
- Triggers: Ideas can spread when people hear frequent reminders of associated ideas. What could be a better trigger than drinking water found to have chemicals in it, or seeing the fence surrounding a hazardous waste site?
- Emotion: Ideas that trigger strong emotion often get shared, particularly when people perceive that information is particularly useful or interesting.
- Public: Humans tend to imitate each other, an idea referred to as "social proof". This phenomenon makes highly visible issues propagate; after all, if so many people are upset about an issue, there must be something to it.
- Practical value: Many people like to help others, so they spread information about ideas that they think people need to know to live healthier, better lives.
- Stories: We humans seem wired to tell stories to convey information. A story about a feared chemical can weave together the other factors that make an idea contagious to make the news compelling. "Ideas that contain social currency and are triggered, emotional, public, practically valuable, and wrapped into stories" can spread like wildfire [137]. Those principles can explain why outrage over an emerging contaminant can explode into wide consciousness, triggering demands for action. We can apply these insights to understanding why some chemical compounds emerge as contaminants and others do not, and potentially gain insight into why some chemical compounds erupt into public view so violently that regulation and scientific research cannot keep pace. In some cases, policy reflects outrage rather than a firm scientific basis.

Six case studies, illustrated in three sets of comparisons, explore this point. These case studies largely pertain to issues discussed elsewhere in this book: perchlorate; DCPA; PFAS; 1, 4-dioxane; and contamination by various persistent organic pollutants in the Arctic. The case studies also include microbeads, a pollutant that emerged rapidly as a concern in the 2010s.

The first comparison contrasts two substances included in the monitoring of public water supplies under UCMR1 between 2001 and 2003. Perchlorate emerged as an environmental contaminant, while DCPA, detected at a similar number of potable water supplies as perchlorate, did not. As a thought experiment and based on generally available information, the outrage factors discussed above were mapped for both perchlorate and DCPA. Figure 3.8 shows

Type of Factor	Risk Component	Perchlorate	DCPA	PFAS	1,4-Dioxane	Arctic Pollution	Microbeads
Select Sandman Outrage Factors	Voluntary vs Coerced	Coerced	Voluntary	Coerced	Coerced	Voluntary	Voluntary
	Natural vs Artifical	Artificial	Artificial	Artificial	Artificial	Artificial	Artificial
	Familiar vs Foreign	Foreign	Familiar	Foreign	Familiar	Foreign	Foreign
	Memorable vs Forgettable	Memorable	Forgettable	Memorable	Forgettable	Forgettable	Memorable
	Chronic vs Catastrophic	Catastrophic	Chronic	Chronic	Chronic	Chronic	Chronic
	Known vs Unknown	Unknown	Known	Unknown	Known	Unknown	Unknown
	Personal Control	No	No to Neutral	No	No	Yes	Yes
	Fairness	No	Yes	No	No	Yes	Yes
	Trustworthy	No	Yes	No	No	Yes	No
Berger Contagious Idea Factors	Social Currency	No	No	Yes	No	Yes	Yes
	Idea Trigger	Yes	Yes	Yes	Yes	No	Yes
	Emotional	No	Yes	Yes	No	No	Yes
	Public	Yes	No	Yes	Yes	No	Yes
	Practical Value	No	Yes	Yes	No	No	Yes
	Stories	No	Yes	Yes	No	Yes	Yes
Outrage?		Yes	No	Yes	Yes	No	Yes
Contagious?		No	Yes	Yes	No	No	Yes

Denotes factor that supports/fuels outrage or contagious idea
Denotes factor that is less likely to fuel outrage or contagious idea

FIGURE 3.8 Thought Experiment on Outrage and Contagion

the results in terms of the likelihood or favorability of outrage emerging and spreading.

Perchlorate first came strongly into public awarenesss following the PEPCON chemical plant explosion in 1988 [138,139]. Two people died in the explosion and over 370 suffered injuries. The explosions damaged buildings within a 1.5-mile radius of the plant, and shock waves extended up to 7 miles away. As a result of this unfortunate incident, many people thought of perchlorate as both memorable and catastrophic. Both factors trigger public outrage. Because of the explosion and ensuing detection in public water supplies during UCMR1, many people doubtless perceived exposure to perchlorate as outside the public's control and felt coerced. Some may have thought it unfair that aerospace and defense companies caused this contamination. One headline in 2011, for example, read "EPA Decides to Limit Rocket Fuel in Drinking Water – Guess Who Objects? Why moms and enviros are happy, and the Perchlorate Information Bureau, i.e. Aerojet, is outraged" [140]. In short, perchlorate may have triggered many public outrage factors, which resulted in this chemical emerging as a compound of environmental concern.

In contrast, people may find DCPA familiar as a chemical available for voluntary use in their gardens. It has other uses that could be positively perceived, such as turf application and protection of food crops. So while detected in similar numbers of drinking water systems as perchlorate, DCPA degradates didn't provoke the same reaction as perchlorate. Short of some public communications regarding DCPA degradates in drinking water in California and some NGO posts, contamination does not appear to have been brought to public attention. As of mid-2019, there isn't any memorable incident surrounding DCPA that has occurred to move it into people's awareness. As a result, DCPA appears to trigger a relatively low number of outrage factors.

UCMR1 and the emergence of perchlorate as an issue occurred in the past. Would these same reactions occur today? That's an interesting question, and to evaluate that further, we looked at the factors of a "contagious idea" as discussed above. Figure 3.8 also shows these factors and contrasts perchlorate and DCPA in a continuation of the thought experiment. Today, knowing about perchlorate (which has been in the public's awareness since the 1980s), doesn't really provide social currency, and it doesn't (today) provide much in terms of a high emotional response. However, the US EPA published the results of UCMR1 around the time when social media was coming into play. As shown on Figure 3.9, there was a lot of public interest surrounding perchlorate in the 2004 to 2006 timeframe as evidenced by the Google searches on this topic [141].

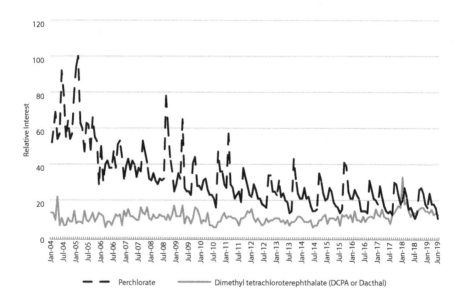

FIGURE 3.9 Relative Interest in Perchlorate versus DCPA as Measured by Searches on Google (US data)

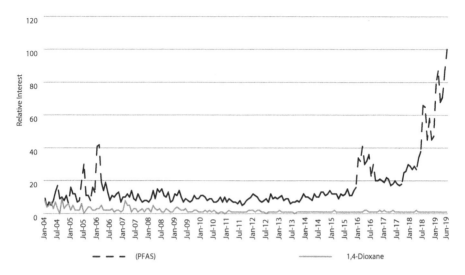

FIGURE 3.10 Relative Interest in PFAS versus 1,4-Dioxane as Measured by Searches on Google (US data)

Social media was relatively new. Facebook use began in earnest circa 2004, YouTube was established in 2005, and Twitter wasn't founded until 2006 [142]. In 2005, social media penetration was just 5% in the US [143] while by 2011 social media had penetrated nearly 50% of the US population with 77% of the population on the internet [144]. Social media presence and connectivity is undoubtedly much higher today and will continue to climb. It is interesting to speculate how social media, if available, would have shaped the public reaction to perchlorate and other contaminants that emerged in the past. Perhaps outrage over perchlorate would have emerged more forcefully than it did.

In this thought experiment, DCPA degradates as an environmental contaminant appear to meet several of the criteria that make an idea contagious. Despite that, outrage has not erupted. As illustrated in Figure 3.9, public interest (as evidenced by Google searches [145]) has remained relatively low.

A second thought experiment pertains to two compounds in the UCMR3 program between 2013 and 2015. Those data showed that 6.9% of public water supply systems sampled contained 1,4-dioxane above 0.35μg/L, its reference concentration (based on the 10^{-6} cancer risk level). Only 0.9% and 0.3% of the public water supplies sampled contained PFOS or PFOA, respectively, above their reference concentrations representing a 10^{-6} lifetime cancer risk (0.07μg/L for both compounds) [146]. However, since the results have been released, PFAS has powerfully emerged as an environmental chemical of concern.

Why has the response to these two contaminants differed? Looking at Figure 3.8 again, PFAS may trigger more outrage factors compared to 1,4-dioxane, which has been in the public's awareness for quite some time. The picture

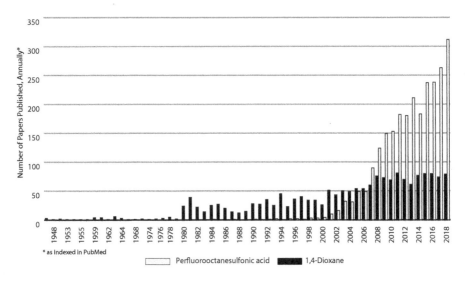

FIGURE 3.11   Scientific Publications on 1,4-Dioxane and PFOS, as indexed in PubMed

becomes clearer when considering the factors that make an idea contagious. These factors seem to resonate far more strongly for PFAS than for 1,4-dioxane. Google search trends provide one measure of public interest in a topic. As shown in Figure 3.10, the frequency of searches for information on PFAS [147] far outweighed that of 1,4-dioxane [148], with interest continuing to build as of June 2019.

A related factor is the number of academic publications on these topics on an annual basis (Figure 3.11). While not a direct measure of public interest, the number of scientific publications shows not only scientific interest but the availability of funding to support scientific research. The conclusions from the research often surface in news articles or blogs, which can generate additional public interest. As shown in Figure 3.11, PFAS (specifically the PFAS chemical PFOS in this case) appears to have generated more interest from a funding and academic standpoint when compared to 1,4-dioxane [149,150].

These data support the inference that PFAS emerged as a contaminant of concern because of public outrage. The rapidity and magnitude of that response likely reflect the factors that make an idea contagious and rapidly spread on social media. As a result, although there are more and more scientific publications on PFAS, the speed of public interest and outrage in PFAS is occurring faster than the scientific community can come to consensus on what is "safe", leading to confusion and unease with the general public. For example, in November 2015, the US EPA issued a provisional health advisory of 400ng/L (400 parts per trillion) for PFOA in drinking water [151]. Six months later, in

May 2016, the EPA issued a PFOA drinking water health advisory of 70ng/L [152]. The draft toxicological profile for PFAS was not released until June 2018 [153], two years later. Public perception of this sequence of events flared into outrage. As expressed in June 2018 [154]:

> A coalition of more than fifty public interest organisations is calling for the release of a "suppressed" assessment of perfluorinated chemicals (PFASs) from the US's Agency for Toxic Substances and Disease Registry (ATSDR). ... It comes following weeks of public outrage over news that the toxicological profile's release has been allegedly slowed. According to internal EPA documents released under a Freedom of Information Act (FOIA) request, the ATSDR assessment will propose safe exposure levels for PFAS chemicals significantly below EPA's non-enforceable drinking water guidelines. But the publicised EPA emails reflect concern at the "public relations nightmare" that this could have. The co-signing NGOs – including the Environmental Working Group (EWG), Greenpeace, Environmental Defense Fund (EDF), and Safer States – are urging the HHS to release the report "immediately". "PFAS chemicals are potent toxicants linked to cancer, liver and thyroid damage, developmental impacts, and numerous other adverse health effects" wrote the groups. "The government should be sharing information about these dangers, not hiding it." The letter also notes an "absence of meaningful action" from the EPA on the substances.

Outrage and the evolution of regulatory requirements for PFAS continue as of this writing in June 2019.

We are not seeing this same phenomenon of outrage outpacing scientific certainty with 1,4-dioxane, per Figure 3.11. Scientists had been publishing papers on 1,4-dioxane for decades and US EPA had already published an IRIS risk assessment prior to the UCMR3 results being published. Therefore, when the public asked about contamination in water supplies – "What does this mean?" – there was an answer. In contrast, when the public questioned whether exposure to PFAS in drinking water presented a significant risk, no clear and definitive answer was available. When regulators and scientists cannot answer, or keep changing the answer, the response further fuels distrust and corresponding outrage, and further propagates the contagious spread of the fear and distrust through social media.

The final comparison contrasts far-away pollution in the Arctic with microbeads in consumer products that, quite literally, bring issues of emerging contaminants home for many people. Section 3.1 included a brief discussion of emerging Arctic contaminants. Microbeads have not been discussed previously in this book and so a short history of their emergence as a contaminant of concerns follows.

Microbeads, used in personal care products such as body washes since the early 1970s, emerged as a contaminant of concern in the 2010s. Concerns about microbeads followed a familiar pattern: while scientists expressed initial concerns about the environmental consequences of washing the tiny plastic beads down the drain in 1991[155], microbeads did not emerge as a contaminant of concern until two decades later. The Ban the Bead campaign began in 2012,

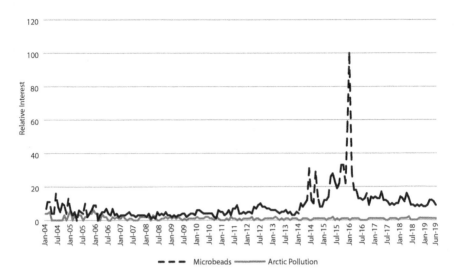

FIGURE 3.12 Relative Interest in Arctic Pollution and Microbeads as Measured by Searches on Google (US data)

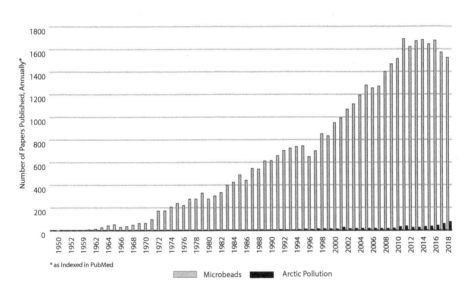

FIGURE 3.13 Publications on Arctic Pollution versus Microbeads over Time, as Indexed in PubMed

after Dutch activists sounded the alarm in 2011. Activists posted pictures of plastic strewn across beaches, floating in ocean gyres and within eviscerated fish that created a strong emotional response. Outrage ignited quickly and outpaced scientific study. Researchers published the first study showing microbeads in surface water in 2013 [156]. In response to public outrage, by 2013 major manu-facturers of personal care products committed to phasing out the use of microbeads. Despite those voluntary actions and the limited data on microbeads in the environment and the effects, regulatory bans soon spread throughout the United States and around the world. The fervency of the calls to ban microbe-ads contrasts with their contribution to plastics pollution in the environment. One estimate suggests that microbeads comprise a small fraction (0.1 to 1.5%) of the plastic found in the ocean today [157]. Microbeads from personal care products are estimated to have contributed 2% of the microplastics pollution in the ocean, according to another study. In comparison, synthetic textiles have contributed 35% of the mass of microplastics and tires another 28% [158]. Clearly, microbeads stirred deep outrage and many of the factors that cause an idea to be contagious (Figure 3.8). That interest is reflected in trends in internet searches and in the growth of academic publications shown in Figures 3.12 [159,160] and 3.13 [161,162].

In contrast, Arctic pollution, arguably presenting a bigger risk, has hardly registered in both the Google Trend data and in academic publications (Fig-ures 3.12 and 3.13). Why is this? When looking at the outrage and contagious factors, Arctic pollution would not seem to provoke as much response. Con-sider the factors shown in Figure 3.8: it is too far away and too remote of an issue to capture broad attention. And there's no simple solution on the order of "Ban the Bead". In short, Arctic pollution may simply be too intangible for outrage and social media contagion to take root.

The emergence of a chemical as an environmental contaminant of concern can happen quickly after a triggering action ignites interest. Analysis of the key principles of outrage and the concept of a contagious idea can help to predict whether a particular compound will emerge. Tracking public interest – through the simple tool of monitoring the volume of internet searches or more sophisticated data mining of blogs, tweets, and other forms of social media – can show how quickly an issue is emerging and thereby inform stra-tegic response. As scientists, we cannot rely solely on our understanding of sci-ence in order to anticipate developments in emerging contaminants. We must incorporate the social aspects of public perception into our evaluations so we can be better prepared to respond to the PFAS of the future.

3.4 THE COMBUSTIBLE MIXTURE: EXPOSURE, HAZARD, AND OUTRAGE

This chapter examined three factors that can combine to ignite the emergence of a "new" contaminant: awareness of exposure; potential for significant, sometimes previously unrecognized, hazard; and public outrage at the possible conse-quences. The contaminants currently in focus often reflect assessments made dec-ades ago and fossilized in regulatory lists. However, our society makes and uses tens of thousands more chemicals than are on those regulatory lists. Only some of the chemicals in commerce represent exposure or hazard concerns. Others, some of which are widely used or dispersed, may have limited hazard character-ization. The evolution of toxicology test methods, data interpretation, and regula-tory requirements for testing can result in the recognition of new hazards that can cause a contaminant to emerge. But as the case studies in this book illustrate, knowledge of exposure or hazard does not always result in the emergence of a contaminant as a concern. Public outrage, which can spread rapidly via social media, can account for the emergence of many environmental concerns that lead swiftly to regulatory action.

NOTES

1. This section is adapted from Section 2.1.2.3 of *Product Stewardship, Life Cycle Analysis and the* Environment (Taylor & Francis /CRC Press, 2015) and is used with permission.
2. Figure 3.3 is adapted from Figure 2.1 of *Product Stewardship, Life Cycle Analysis and the Environment* (Taylor & Francis/CRC Press, 2015) and is used with permission.
3. Reprinted with permission from the National Academies Press, Copyright 2009, National Academy of Sciences.
4. This section is adapted from Section 2.1.2.2 of *Product Stewardship, Life Cycle Analysis and the Environment* (Taylor & Francis/CRC Press, 2015) and is used with permission.

REFERENCES

1. US EPA, 2018. IRIS assessments. Web page last updated December 18, 2018. Available at: https://cfpub.epa.gov/ncea/iris_drafts/atoz.cfm?list_type=alpha (accessed January 31, 2019).
2. Massachusetts Contingency Plan. 310 CMR 40.0974, Table 1. Available at: www.mass.gov/files/documents/2017/10/17/310cmr40.pdf (accessed January 31, 2019).
3. US EPA, 2019. EPA releases first major update to chemicals list in 40 years. Press release February 19, 2019. Available at: www.epa.gov/newsreleases/epa-releases-first-major-update-chemicals-list-40-years (accessed May 1, 2019).
4. 42 U.S. Code § 9601. Definitions.
5. UNEP, 2019. The evolving chemicals economy: Status and trends relevant for sus-tainability. Global chemicals outlook II part I. Chapter 5. SAICM/OEWG.3/INF/25 (Part I). Available at: http://wedocs.unep.org/handle/20.500.11822/28186 (accessed June 1, 2019).
6. Bajak, F., & Olsen, L., 2019. Silent spills: Environmental damage from Hurricane Harvey is just beginning to emerge. *Houston Chronicle*. March 5.
7. US EPA, 2016. Methyl Tertiary Butyl Ether (MTBE) – Gasoline. Web page last updated February 20, 2016. Available at: https://archive.epa.gov/mtbe/web/html/gas.html (accessed June 12, 2019).
8. Lorenzetti, M.S. 1994. On the road with oxygenates. *Chemical Business* (January): 15–17. As cited in: ATSDR, 1996. Toxicological Profile for Methyl tert-Butyl Ether. Available at: www.atsdr.cdc.gov/ToxProfiles/tp91.pdf (accessed June 12, 2019).
9. US EPA, 2016. Methyl Tertiary Butyl Ether (MTBE) – Drinking water. Web page last updated February 20, 2016. Available at: https://archive.epa.gov/mtbe/web/html/water.html (accessed June 12, 2019).

10. ATSDR, 1996. Toxicological profile for methyl tert-butyl ether. Available at: www.atsdr.cdc.gov/ToxProfiles/tp91.pdf (accessed June 12, 2019).

11. US EPA, 2008. The analysis of occurrence data from the first unregulated contaminant monitoring regulation (UCMR 1) in support of regulatory determinations for the second drinking water contaminant candidate list (CCL 2). EPA 815-R-08-013. Available at: www.epa.gov/sites/production/files/2014-09/documents/the_analysis_of_occurrence_data_from_ucmr1_in_support_of_rd2.pdf (accessed June 12, 2019).

12. US EPA, 2007. State actions banning MTBE (statewide). EPA420-B-07-013 August 2007.

13. Sweet, F., Kauffman, M., Pellerin, T., Espy, D., & Mills, M. 2005. An estimate of the national cost for remediation of MTBE releases from existing leaking underground storage tank sites. ENSR International. July 2005. Available at: https://clu-in.org/download/contaminantfocus/mtbe/MTBEReport.pdf (accessed June 12, 2019).

14. US EPA, 2019. ChemView: T-Butyl methyl ether, 1634- 04-4. Available at: https://chemview.epa.gov/chemview# (accessed June 12, 2019).

15. US EPA, 2019. TRI-listed chemicals. Available at: www.epa.gov/toxics-release-inventory-tri-program/tri-listed-chemicals (accessed April 29, 2019).

16. US EPA, 2019. "TRI national analysis" releases of chemicals. Available at: www.epa.gov/trinationalanalysis/releases-chemicals (accessed April 29, 2019).

17. Prevedouros, K., Cousins, I.T., Buck, R.C., & Korzeniowski, S.H. 2006. Sources, fate and transport of perfluorocarboxylates. *Environmental Science & Technology*, 40(1), pp. 32–44.

18. Lang, J.R., Allred, B.M., Field, J.A., Levis, J.W., & Barlaz, M.A. 2017. National estimate of per-and polyfluoroalkyl substance (PFAS) release to US municipal landfill leachate. *Environmental Science & Technology*, 51(4), pp. 2197–2205.

19. ATSDR, 2008. Toxicological profile for perchlorates. Draft for public comment. Agency for Toxic Substances and Disease Registry. September.

20. US Department of the Navy, Office of Naval Research. Undated. Science and technology focus: Ocean in motion: Currents – Characteristics. Available at: www.onr.navy.mil/focus/ocean/motion/currents1.htm (accessed February 23, 2014).

21. US National Oceanographic and Atmospheric Administration, 2007. NOAA's National Ocean Service: Education Currents: Waves. Available at: http://oceanservice.noaa.gov/education/tutorial_currents/lessons/currents_tutorial.pdf (accessed February 9, 2014).

22. Broecker, W.S. 1991. The great ocean conveyor. *Oceanography*, 4(2), pp. 79–89.

23. Ross, D. 1995. *Introduction to Oceanography.* New York: HarperCollins College Publishers, pp. 199–226, 339–343. As cited by National Oceanographic and Atmospheric Administration, 2007. NOAA's National Ocean Service: Education Currents: Waves. Available at: http://oceanservice.noaa.gov/education/tutorial_currents/lessons/currents_tutorial.pdf (accessed February 9, 2014).

24. Zarfl, C., Scheringer, M., & Matthies, M. 2011. Screening criteria for long-range transport potential of organic substances in water. *Environmental Science & Technology*, 45, pp. 10075–10081.

25. United Nations Environmental Programme, 2006. Report of the persistent organic pollutants Review Committee on the work of its second meeting. Addendum: Risk profile on perfluorooctane sulfonate. Available at: http://chm.pops.int/TheConvention/ThePOPs/TheNewPOPs/tabid/2511/Default.aspx (accessed February 16, 2013).

26. Sellers, K., 2015. Product stewardship, life cycle analysis and the environment. Table 2.2. Taylor & Francis/CRC Press.

27. Ahrens, L., Felizeter, S., Sturm, R., Xie, Z., & Ebinghaus, R. 2009. Polyfluorinated compounds in waste water treatment plant effluents and surface waters along the River Elbe. *Germany. Marine Pollution Bulletin*, 58, pp. 1326–1333.

28. Ahrens, L., Gerwinski, W., Theobald, N., & Ebinghaus, A. 2010. Sources of poly-fluoroalkyl compounds in the North Sea, Baltic Sea and Norwegian Sea: Evidence from their spatial distribution in surface water. *Marine Pollution Bulletin*, 60, pp. 255–260.

29. Theobald, N., Caliebe, C., Gerwinski, W., Huhnerfuss, H., & Lepom, P. 2011. Occurrence of perfluorinated organic acids in the North and Baltic seas. Part 1: Distribution in sea water. *Environmental Science and Pollution Research*, 18(7), pp. 1057–1069.

30. Ahrens, L., Barber, J.L., Xie, Z., & Ebinghaus, A. 2009. Longitudinal and latitudinal distribution of perfluoroalkyl compounds in the surface water of the Atlantic Ocean. *Environmental Science & Technology*, 43, pp. 3122–3127.

31. Ahrens, L., Xie, Z., & Ebinghaus, R. 2010. Distribution of perfluoroalkyl compounds in seawater from Northern Europe, Atlantic Ocean, and Southern Ocean. *Chemosphere*, 78(8), pp. 1011–1016.

32. Cai, M., Zhao, Z., Yin, Z., Ahrens, L., Huang, P., Cai, M., Yang, H., He, J., Sturm, R., Ebinghaus, R., & Xie, Z., 2012. Occurrence of perfluoroalkyl compounds in surface waters from the North Pacific to the Arctic Ocean. *Environmental Science & Technology*, 46(2), pp. 661–668.

33. Cai, M., Zhao, Z., Yang, H., Yin, Z., Hong, Q., Sturm, R., Ebinghaus, R., Ahrens, L., Cai, M., He, J., & Xie, Z. 2012. Spatial distribution of per-and poly-fluoroalkyl compounds in coastal waters from the East to South China Sea. *Environmental Pollution*, 161, pp. 162–169.

34. Yamashita, N., Taniyasu, S., Petrick, G., Wei, S., Gamo, T., Lam, P.K., & Kannan, K. 2008. Perfluorinated acids as novel chemical tracers of global circulation of ocean waters. *Chemosphere*, 70(7), pp. 1247–1255.

35. Committee on the Significance of International Transport of Air Pollutants; National Research Council, 2009. *Global Sources of Local Pollution: An Assessment of Long-Range Transport of Key Air Pollutants to and from the United States*. Washington, DC: The National Academies Press. pp. 1–7; 18–23; 113–121. Appendix B. Available at: www.nap.edu/catalog.php?record_id=12743 (accessed February, 2014).

36. National Geographic Education, 2014. Encyclopedic Entry: Jet Stream. Available at: http://education.nationalgeographic.com/education/encyclopedia/jet-stream/?ar_a=1 (accessed February 9, 2014).

37. Gouin, T., Mackay, D., Jones, K.C., Harner, T., & Meijer, S.N. 2004. Evidence for the "grasshopper" effect and fractionation during long-range atmospheric transport of organic contaminants. *Environmental Pollution*, 128(1–2), pp. 139–148.

38. von Waldow, H., MacLeod, M., Jones, K., Scheringer, M., & Hungerbühler, K. 2010. Remoteness from emission sources explains the fractionation pattern of poly-chlorinated biphenyls in the Northern Hemisphere. *Environmental Science & Technology*, 44(16), pp. 6183–6188.

39. AMAP, 2017. Chemicals of emerging arctic concern summary for policy-makers. Available at: www.amap.no/documents/download/2890/inline (accessed May 31, 2019).

40. US EPA, 2003. Status of EPA's interim assessment guidance for perchlorate. Web page last updated April 2019. Available at: www.epa.gov/fedfac/status-epas-interim-assessment-guidance-perchlorate (accessed June 1, 2019).

41. US National Library of Medicine National Institutes of Health. Results of key word search for "perchlorate". Available at: www.ncbi.nlm.nih.gov/pubmed/?term=Perchlorate (accessed February 17, 2019).

42. Sellers, K., Weeks, K., Alsop, W.R., Clough, S.R., Hoyt, M., Pugh, B., & Robb, J. 2006. *Perchlorate: Environmental Problems and Solutions.* Boca Raton, FL: CRC Press.

43. ATSDR, 2008. *Toxicological Profile for Perchlorates.* US Department of Health and Human Services Public Health Service, Agency for Toxic Substances and Disease Registry. Available at: www.atsdr.cdc.gov/toxprofiles/tp162.pdf (accessed January 31, 2019).

44. Centers for Disease Control, 2016. National Biomonitoring Program. Biomonitoring summary – Perchlorate, CAS No. 7601-90-3.Web page last updated: Tuesday, December 27 2016. Available at: www.cdc.gov/biomonitoring/Perchlorate_Biomoni toringSummary.html (accessed January 31, 2019).

45. US EPA, 2014. Technical fact sheet – Perchlorate. Available at: www.epa.gov/sites/production/files/2014-03/documents/ffrrofactsheet_contaminant_perchlorate_ja nuary2014_final.pdf (accessed February 1, 2019).

46. US EPA, 2008. Interim drinking water health advisory for perchlorate. EPA 822-R-08-025.

47. Cox, E., 2005. Evaluation of alternative causes of wide-spread, low concentration perchlorate impacts to groundwater. Report prepared for the US Strategic Environmental Research and Development Program by Geosyntec Consultants. Boxborough, MA. May.

48. US Patent Office, s.v. "Perchlorate". Available at: www.uspto.gov/ (accessed August 28, 2005). As cited in Sellers, K., Weeks, K., Alsop, W.R., Clough, S.R., Hoyt, M., Pugh, B. and Robb, J., 2006. *Perchlorate: Environmental problems and solutions.* CRC Press.

49. US EPA, 2011. Drinking water: Regulatory determination on perchlorate. *Federal Register,* 76(29) (Friday, February 11, 2011), pp. 7762–7767.

50. USGS, undated. National Water-Quality Assessment (NAWQA). Available at: www.usgs.gov/mission-areas/water-resources/science/national-water-quality-assessment-nawqa?qt-science_center_objects=0#qt-science_center_objects (accessed June 9, 2019).

51. CDC, 2017. National Biomonitoring Program. Web page last updated April 7, 2017. Available at: www.cdc.gov/biomonitoring/index.html (accessed February 17, 2019).

52. US EPA, 2018. Learn about the unregulated contaminant monitoring rule. Web page last updated January 24, 2018. Available at: www.epa.gov/dwucmr/learn-about-unregulated-contaminant-monitoring-rule (accessed June 10, 2019).

53. US EPA, 2019. Monitoring the occurrence of unregulated drinking water contaminants. Web page last updated February 15, 2019. Available at: www.epa.gov/dwucmr (accessed April 19, 2019).

54. Office of the President, 2018. Plan for addressing critical research gaps related to {\hskip 0.7pt}emerging contaminants in drinking water. A Report by the Task Force for Emerging Contaminants of the National Science & Technology Council. October. Available at: www.whitehouse.gov/wp-content/uploads/2018/11/Plan-for-Addressing-Critical-Research-Gaps-Related-to-Emerging-Contaminants-in-Drinking-Water.pdf (accessed April 19, 2019).

55. US EPA, undated. Occurrence data for the unregulated contaminant monitoring rule. UCMR 1 (2001–2005) Occurrence Data: UCMR 1 List 1 and List 2 chemical monitoring data. Available at: www.epa.gov/dwucmr/occurrence-data-unregulated-contaminant-monitoring-rule#1 (accessed June 1, 2019).

56. US EPA, 2008. Regulatory determinations support document for selected contaminants from the second drinking water contaminant candidate list (CCL 2). Chapter 4 (DCPA Mono- and di-acid degradates). EPA Report 815-R-08-012. Available at: www.epa.gov/sites/production/files/2014-09/documents/chapter_4_dcpa_mono-_and_di-acid_degradates.pdf (accessed June 1, 2019).

57. USGS, 2016. Pesticide National Synthesis Project. Estimated annual agricultural pesticide use – Pesticide use maps. Estimated agricultural use for DCPA, 2016 (Preliminary). Available at: https://water.usgs.gov/nawqa/pnsp/usage/maps/show_map. php?year=2016&map=DCPA&hilo=L (accessed June 1, 2019).

58. Lohstroh, P., & Koshlukova, S., 2017. Memorandum to: Shelley DuTeaux, PhD MPH, Chief Human Health Assessment Branch, California Department of Pesticide Registration. Subject: Evaluation of the potential human health effects from drinking ground water containing Dacthal (DCPA) degradates. February 23, 2017. Available at: www.cdpr.ca.gov/docs/hha/memos/tpa%20in%20ground%20water%20reply%20final%2002232017%20complete%20executed.pdf (accessed June 6, 2019).

59. Temkin, A., 2019. More than half of kale samples tainted by possibly cancer-causing pesticide. March 20, 2019. Available at: www.ewg.org/foodnews/kale.php (accessed June 1, 2019).

60. US EPA, 2018. *Chemicals Evaluated for Carcinogenic Potential Annual Cancer Report 2018*. US Environmental Protection Agency, Office of Pesticide Programs. Available at: http://npic.orst.edu/chemicals_evaluated.pdf (accessed June 1, 2019).

61. Sutherland-Ashley, K., Eya, B., & Lim, L., 2018. *Public Health Concentrations for Chlorthal-dimethyl (DCPA) and its Degradates Monomethyl Tetrachloroterephthalic Acid (MTP) and Tetrachloroterepththalic Acid (TPA) in Groundwater*. Pesticide and Environmental Toxicology Branch, Office of Environmental Health Hazard Assessment, California Environmental Protection Agency. Available at: https://oehha.ca.gov/media/downloads/pesticides/report/dcpa083018.pdf (accessed June 1, 2019).

62. OEHHA, 2016. *Perchloroethylene Inhalation Cancer Unit Risk Factor. Technical Support Document for Cancer Potency Factors. Appendix B*. Office of Environmental Health Hazard Assessment. September.

63. US EPA, 2012. *IRIS Chemical Assessment Summary Tetrachloroethylene (Perchloroethylene); CASRN 127-18-4*. Available at: https://cfpub.epa.gov/ncea/iris/iris_documents/documents/subst/0106_summary.pdf (accessed June 11, 2019).

64. US EPA, 2012. Toxicological review of tetrachloroethylene (perchloroethylene). US Environmental Protection Agency. EPA/635/R-08/011F. February.

65. US EPA, 2012. Toxicological review of tetrachloroethylene (perchloroethylene). US Environmental Protection Agency. EPA/635/R-08/011F. February. pp. 3–30.

66. US EPA, 2012. Toxicological review of tetrachloroethylene (perchloroethylene). US Environmental Protection Agency. EPA/635/R-08/011F. February. pp. 3–48.

67. US EPA, 2013. Toxicological review of 1,4-dioxane (with inhalation update). US Environmental Protection Agency. EPA/635/R-11/003F. September. p. 135.

68. Dourson, M., Reichard, J., Nance, P., Burleigh-Flayer, H., Parker, A., Vincent, M., & McConnell, E.E. 2014. Mode of action analysis for liver tumors from oral 1,4-dioxane exposures and evidence-based dose response assessment. *Regulatory Toxicology and Pharmacology*, 68(3), pp. 387–401.

69. Dourson, M.L., Higginbotham, J., Crum, J., Burleigh-Flayer, H., Nance, P., Forsberg, N.D., Lafranconi, M., & Reichard, J. 2017. Update: Mode of action (MOA) for liver tumors induced by oral exposure to 1,4-dioxane. *Regulatory Toxicology and Pharmacology*, 88, pp. 45–55.

70. Becker, R.A., Dellarco, V., Seed, J., Kronenberg, J.M., Meek, B., Foreman, J., Palermo, C., Kirman, C., Linkov, I., Schoeny, R., et al., 2017. Quantitative weight of evidence to assess confidence in potential modes of action. *Regulatory Toxicology and Pharmacology*, 86, pp. 205–220.

71. Health Canada, 2018. *1,4-Dioxane in Drinking Water*. Guideline Technical Document for Public Consultation.

72. Johnson, P.D., Goldberg, S.J., Mays, M.Z., & Dawson, B.V. 2003. Threshold of trichloroethylene contamination in maternal drinking waters affecting fetal heart development in the rat. *Environmental Health Perspectives*, 111(3), pp. 289–292.

73. US EPA, 2014. Compilation of information relating to early/interim actions at Superfund sites and the TCE IRIS assessment. Memorandum from Robin Richardson, Acting Director, Offices of Superfund Remediation and Technology Innovation to Superfund Division Directors, EPA Regions 1–10. August 27.

74. Ohio EPA, 2016. Guidance document. Recommendations regarding response action levels and timeframes for common contaminants of concern at vapor intrusion sites in Ohio. Ohio Environmental Protection Agency. August.

75. Makris, S.L., Scott, C.S., Fox, J., Knudsen, T.B., Hotchkiss, A.K., Arzuaga, X., Euling, S.Y., Powers, C.M., Jinot, J., Hogan, K.A., & Abbott, B.D., 2016. A systematic evaluation of the potential effects of trichloroethylene exposure on cardiac development. *Reproductive Toxicology*, 65, pp. 321–358.

76. Wikoff, D., Urban, J.D., Harvey, S., & Haws, L.C. 2018. Role of risk of bias in systematic review for chemical risk assessment: A case study in understanding the relationship between congenital heart defects and exposures to trichloroethylene. *International Journal of Toxicology*, 37(2), pp. 125–143.

77. Sussan, T.E., Leach, G.J., Covington, T.R., Gearhart, J.M., & Johnson, M.S., 2019. Trichloroethylene occupational exposure level for the Department of Defense. US Army Public Health Center. January.

78. ECHA, undated. Trichloroethylene registration dossier. Toxicological summary. Available at: https://echa.europa.eu/registration-dossier/-/registered-dossier/14485/7/1 (accessed June 11, 2019).

79. 3B2 twb 0.39w?>Regulation (EC) No 1907/2006 of the European Parliament and of the Council of 18 December 2006 concerning the Registration, Evaluation, Authorisation and Restriction of Chemicals (REACH), establishing a European Chemicals Agency, amending Directive 1999/45/EC and repealing Council Regulation (EEC) No 793/93 and Commission Regulation (EC) No 1488/94 as well as Council Directive 76/769/EEC and Commission Directives 91/155/EEC, 93/67/EEC, 93/105/EC and 2000/21/EC (Text with EEA relevance). Available at: http://data.europa.eu/eli/reg/2006/1907/2018-05-09 (accessed June, 2019).

80. ECHA, 2019. Registered substances. Web site last updated June 5, 2019. Available at: https://echa.europa.eu/information-on-chemicals/registered-substances;jsessionid=C8B11FF15E219166EFD8031625A61A03.live2 (accessed June 5, 2019).

81. US EPA, 2019. SNURs, exported from chem view database. Available at: https://java.epa.gov/chemview# (accessed June 2, 2019).

82. OSHA, undated. About OSHA. Available at: www.osha.gov/about.html (accessed June 21, 2019).

83. OSHA, PELs update. Available at: www.osha.gov/archive/oshinfo/priorities/pel.html (accessed June 21, 2019).

84. US EPA, 2014. TSCA work plan for chemical assessments: 2014 update. US Environmental Protection Agency. Office of Pollution Prevention and Toxics. October.

85. US EPA, 2018. *A Working Approach for Identifying Potential Candidate Chemicals for Prioritization*. US Environmental Protection Agency. Office of Chemical Safety

and Pollution Prevention. September. p. 6. Available at: www.epa.gov/sites/produc tion/files/2018-09/documents/preprioritization_white_paper_9272018.pdf (accessed June 21, 2019).

86. US EPA, 2019. Initiation of Prioritization Under the Toxic Substances Control Act (TSCA). Federal Register: March 21, 2019 (Volume 84, Number 55, Page 10491).

87. US EPA, 1999. Category for persistent, bioaccumulative, and toxic new chemical substances. *Federal Register*, November 4, 1999 64(213).

88. US EPA, 2012. Sustainable futures / P2 framework manual 2012. EPA-748-B12 -001. Chapter 7. Estimating persistence, bioaccumulation, and toxicity using the PBT profiler. Available at: www.epa.gov/sites/production/files/2015-05/documents/ 07.pdf (accessed March 11, 2019).

89. ECHA, 2017. Guidance on information requirements and chemical safety assess- ment. Chapter R.11: PBT/vPvB assessment. Version 3.0. June 2017. Table R.11–1. Available at: https://echa.europa.eu/documents/10162/13632/information_require ments_r11_en.pdf (accessed April 16, 2019).

90. United Nations Environmental Programme, 2017. Stockholm convention on per- sistent organic pollutions (POPs), text and annexes (revised in 2017). Annex D: Information Requirements and Screening Criteria. Available at: http://chm .pops.int/TheConvention/Overview/TextoftheConvention/tabid/2232/Default.aspx (accessed May 31, 2019).

91. US EPA, 2013. Interpretive assistance document for assessment of discrete organic chemicals sustainable futures summary assessment. Page 5. Available at: www .epa.gov/sites/production/files/2015-05/documents/05-iad_discretes_june2013.pdf (accessed March 11, 2019).

92. Klecka, G.M., & Muir, D.C.G., 2008. Science-based guidance and framework for the evaluation and identification of PBTs and POPs: Summary of a SETAC Pell- ston workshop. SETAC Pellston Workshop on Science-Based Guidance and Framework for the Evaluation and Identification of PBTs and POPs; 2008 Jan 28– Feb 1; Pensacola Beach, FL: Society of Environmental Toxicology and Chemistry (SETAC). Available at: https://setac.onlinelibrary.wiley.com/doi/10.1897 /IEAM_2009-045.1 (accessed April 16, 2019).

93. Scow, K.M., & Johnson, C.R., 1997. Effect of sorption on biodegradation of soil pollutants. *Advances in Agron*, 58, pp. 1–56.

94. Sinkkonen, S., & Paasivirta, J. 2000. Degradation half-life times of PCDDs, PCDFs and PCBs for environmental fate modeling. *Chemosphere*, 40(9–11), pp. 943–949.

95. US Geological Survey, 2013. Bioaccumulation – Definitions. Web page last updated July 10, 2013. Available at: http://toxics.usgs.gov/definitions/bioaccumula tion.html (accessed January 12, 2014).

96. US EPA, 2012. User's guide and technical documentation – KABAM version 1.0. (Kow (based) aquatic bioaccumulation model). Appendix F. Description of equa- tions used to calculate the BCF, BAF, BMF, and BSAF values. Web page last updated May 9, 2012. Available at: www.epa.gov/oppefed1/models/water/kabam/ kabam_user_guide_appendix_f.html (accessed January 12, 2014).

97. US Geological Survey, 2013. Bioconcentration – Definitions. Web page last updated July 10, 2013. Available at: http://toxics.usgs.gov/definitions/bioconcentra tion.html (accessed January 12, 2014).

98. Hu, H., Xu, F., Li, B., Cao, J., Dawson, R., & Tao, S. 2005. Prediction of the bio- concentration factor of PCBs in fish using the molecular connectivity index and fragment constant models. *Water Environment Research*, 77(1), pp. 87–97.

99. Strempel, S., Scheringer, M., Ng, C.A., & Hungerbühler, K. 2012. Screening for PBT chemicals among the "existing" and "new" chemicals of the EU. *Environmental Science & Technology*, 46(11), pp. 5680–5687.

100. US EPA, 2017. Preliminary information on manufacturing, processing, distribution, use, and disposal: Pentachlorothiophenol. Available at: www.epa.gov/sites/pro duction/files/2017-08/documents/pctp_-_use_information-8-7-17-v3-clean.pdf (accessed June 6, 2019).

101. US EPA, 2018. Exposure and use assessment of five persistent, bioaccumulative and toxic chemicals peer review draft. EPA Document # EPA-740-R1-8002. Available at: www.epa.gov/sites/production/files/2018-06/documents/exposure_use_as sessment_five_pbt_chemicals.pdf (accessed June 6, 2019).

102. To-Figueras, J., Gómez-Catalán, J., Rodamilans, M., & Corbella, J. 1992. Sulphur derivative of hexachlorobenzene in human urine. *Human & Experimental Toxicology*, 11(4), pp. 271–273.

103. To-Figueras, J., Barrot, C., Sala, M., Otero, R., Silva, M., Ozalla, M.D., Herrero, C., Corbella, J., Grimalt, J., & Sunyer, J. 2000. Excretion of hexachlorobenzene and metabolites in feces in a highly exposed human population. *Environmental Health Perspectives*, 108(7), pp. 595–598.

104. US EPA, undated. Chemview entry for pentachlorothiophenol, CAS 133-49-3. Available at: https://chemview.epa.gov/chemview (accessed June 21, 2019).

105. Reemtsma, T., Berger, U., Arp, H.P.H., Gallard, H., Knepper, T.P., Neumann, M., Quintana, J.B., & Voogt, P.D. 2016. Mind the gap: Persistent and mobile organic compounds – Water contaminants that slip through. *Environmental Science & Technology*, Oct 4, 50(19), pp. 10308–10315.

106. German Environment Agency, 2017. Protecting the sources of our drinking water: A revised proposal for implementing criteria and an assessment procedure to identify Persistent, Mobile and Toxic (PMT) and very Persistent, very Mobile (vPvM) substances registered under REACH. Available at: www.umweltbundesamt.de /sites/default/files/medien/1410/publikationen/171027_uba_pos_pmt_substances_en gl_2aufl_bf.pdf (accessed April 16, 2019).

107. SCHEER (Scientific Committee on Health, Environmental and Emerging Risks), 2018. Statement on emerging health and environmental issues. 20 December 2018. Available at: https://ec.europa.eu/health/sites/health/files/scientific_committees/ scheer/docs/scheer_s_002.pdf (accessed April 16, 2019).

108. Davies, E., 2017. German environment agency updates criteria for "mobile" chemicals in water. 19 October. Chemical Watch. Available at: https://chemicalwatch .com/60218/german-environment-agency-updates-criteria-for-mobile-chemicals-in- water (accessed April 16, 2019).

109. Arp, H.P.H., Brown, T.N., Berger, U., & Hale, S.E. 2017. Ranking REACH registered neutral, ionizable and ionic organic chemicals based on their aquatic persistency and mobility. *Environmental Science: Processes & Impacts*, 19(7), pp. 939–955.

110. Schulze, S., Sättler, D., Neumann, M., Arp, H.P.H., Reemtsma, T., & Berger, U. 2018. Using REACH registration data to rank the environmental emission potential of persistent and mobile organic chemicals. *Science of the Total Environment*, 625, pp. 1122–1128.

111. The Norwegian Geotechnical Institute, 2018. Preliminary assessment of substances registered under REACH that could fulfil the proposed PMT/vPvM criteria. Background document to the workshop: "PMT and vPvM substances under REACH" March 13th–14th. Bundespresseamt, Berlin, Germany. DOC.NO. 20160426-TN-01 2018-03-06. Available at: www.umweltbundesamt.de/en/2018-workshop-pmt- substances-under-reach (accessed April 16, 2019).

112. Montes, R., Aguirre, J., Vidal, X., Rodil, R., Cela, R., & Quintana, J.B. 2017. Screening for polar chemicals in water by trifunctional mixed-mode liquid chromatography–high resolution mass spectrometry. *Environmental Science & Technology*, 51(11), pp. 6250–6259.

113. Rosen, J.D. 1990. Much ado about alar. *Issues in Science and Technology*, 7(1), pp. 85–90.

114. American Council on Science and Health, 1999. An unhappy anniversary: The alar "scare" ten years later. Available at: www.acsh.org/news/1999/02/01/an-unhappy-anniversary-the-alar-scare-ten-years-later (accessed May 2, 2019).

115. Mendes, G.C., Brandao, T.R., & Silva, C.L. 2007. Ethylene oxide sterilization of medical devices: A review. *American Journal of Infection Control*, 35(9), pp. 574–581.

116. National Institute for Occupational Safety and Health, 1981. Ethylene oxide (EtO): Evidence of carcinogenicity. US Department of Health and Human Services, Public Health Service, Centers for Disease Control, National Institute for Occupational Safety and Health, Robert A. Taft Laboratories.

117. ChemView, 2019. Manufacturing, processing, use and release data for ethylene oxide 75-21-8, chemical data reporting summary on national aggregate production volumes. Retrieved May 28, 2019.

118. The Right to Know Network, 2019. TRI summary for ethylene oxide 75-21-8 for reporting year 2016. Retrieved May 28, 2019.

119. Hawthorne, M., 2018. Officials knew ethylene oxide was linked to cancer for decades. Here's why it's still being emitted in Willowbrook and Waukegan. *Chicago Tribune*, December 20, 2018. Retrieved on May 20, 2019.

120. Baichwal, R., 2018. Illinois EPA orders Sterigenics to cease operations in Willowbrook until safety review complete. Abc7chicago.com. October 2, 2018. Retrieved on May 20, 2019.

121. ABC7, 2019. Willowbrook Sterigenics plan shut down Friday night. Abc7chicago.com. February 15, 2019. Retrieved on May 20, 2019.

122. Food and Drug Administration, 2019. Statement from FDA Commissioner Scott Gottlieb, M.D., on steps the agency is taking to prevent potential medical device shortages and ensure safe and effective sterilization amid shutdown of a large contract sterilization facility. *FDA Statement*. March 26, 2019.

123. Food and Drug Administration, 2019. Statement from Jeff Shuren, M.D., director of the Center for Devices and Radiological Health, on agency efforts to mitigate temporary shortage of pediatric breathing tubes due to recent closure of Illinois sterilization facility. *FDA Statement*. April 12, 2019.

124. Carlson, J., 2019. Sterilization plant closure in Illinois is a challenge for Minnesota medical device makers. *Star Tribune*, April 13, 2019. Retrieved on May 20, 2019.

125. Sandman, P., 1993. Responding to community outrage. Fairfax: American Industrial Hygiene Association. Available at: www.psandman.com/media/RespondingtoCommunityOutrage.pdf (accessed January 1, 2019).

126. Freakonomics, 2011. Risk = Hazard + Outrage: A conversation with risk consultant Peter Sandman. *Freakonomics*, November 29, 2011. Available at: http://freakonomics.com/2011/11/29/risk-hazard-outrage-a-conversation-with-risk-consultant-peter-sandman/ (accessed March 28, 2019).

127. The Nobel Prize in physiology or medicine 1948. NobelPrize.org. Nobel Media AB 2019. Sun. 3 Feb 2019. Available at: www.nobelprize.org/prizes/medicine/1948/summary/ (accessed February 3, 2019).

128. Mueller, P., 1948. Award ceremony speech. NobelPrize.org. Nobel Media AB 2019. Available at: www.nobelprize.org/prizes/medicine/1948/ceremony-speech/ (accessed February 3, 2019).

129. Blood, N.O.F. 1948. Pesticides: Chemical contaminants of foods. *JAMA*, 137(18), pp. 1604–1605.

130. US Tariff Commission, 1948. Synthetic organic chemicals United States Production and Sales, 1948 GPO CI. No. 164 TC 1.9:1641 second series.

131. US Tariff Commission, 1962. Synthetic organic chemicals United States Production and Sales, 1962 TC publication 114.

132. Carson, R. 2002. *Silent Spring*. New York, NY: Houghton Mifflin. Anniversary Edition, originally published 1962.

133. Griswold, E., 2012. How "Silent Spring" ignited the environmental movement. *New York Times*. September 21. Available at: www.nytimes.com/2012/09/23/magazine/how-silent-spring-ignited-the-environmental-movement.html (accessed June, 2019).

134. Lear, L. 2019. The life and legacy of Rachel Carlson, Silent Spring. Available at: www.rachelcarson.org/SilentSpring.aspx (accessed June 1, 2019).

135. US EPA, 1972. DDT Ban takes effect [EPA press release – December 31, 1972]. Available at: https://archive.epa.gov/epa/aboutepa/ddt-ban-takes-effect.html (accessed June 1, 2019).

136. Vosoughi, S., Roy, D., & Aral, S. 2018. The spread of true and false news online. *Science*, 359(6380), pp. 1146–1151.

137. Berger, J. 2013. *Contagious: Why Things Catch On*. New York: Simon & Schuster.

138. Mniszewski, K.R. 1994. The PEPCON plant fire/explosion: A rare opportunity in fire/explosion investigation. *Journal of Fire Protection Engineering*, 6(2), pp. 63–78.

139. Routley, J.G., Fire and explosions at rocket fuel plant, Henderson Nevada (May 4, 1988). Report 021 – Major Fires Investigation Project. Federal Emergency Management Agency, United States Fire Administration National Fire Data Center. 1988.

140. Weinstein, D., 2011. EPA decides to limit rocket fuel in drinking water – Guess who objects? Why moms and enviros are happy, and the Perchlorate Information Bureau, i.e. Aerojet, is outraged. *The Atlantic*. February 3. Available at: www.theatlantic.com/politics/archive/2011/02/epa-decides-to-limit-rocket-fuel-in-drinking-water-guess-who-objects/342385/ (accessed June, 2019).

141. Google Trends Public Data, 2019. Keyword search for "Perchlorate". Available at: https://trends.google.com/trends/explore?date=all&geo=US&q=perchlorate (accessed June 10, 2019).

142. Edosomwan, S., Prakasan, S.K., Kouame, D., Watson, J., & Seymour, T., 2011. The history of social media and its impact on business. *Journal of Applied Management and Entrepreneurship*, 16(3), pp. 79–91.

143. Terrell, 2019. The history of social media: Social networking evolution! *History Cooperative*. Available at: https://historycooperative.org/the-history-of-social-media/ (accessed June 13, 2019).

144. Thompson, R., 2011. Radicalization and the use of social media. *Journal of Strategic Security*, 4(4), pp. 167–190.

145. Google Trends Public Data, 2019. Keyword search for "DCPA + Dacthal". Available at: https://trends.google.com/trends/explore?date=all&geo=US&q=DCPA%20%2B%20Dacthal (accessed June 10, 2019).

146. US EPA, 2017. The third unregulated contaminant monitoring rule (UCMR3): Data summary, January 2017. Available at: www.epa.gov/sites/production/files/2017-02/documents/ucmr3-data-summary-january-2017.pdf (accessed June, 2019).

147. Google Trends Public Data, 2019. Keyword search for "PFOS + PFOA + PFAS". Available at: https://trends.google.com/trends/explore?date=all&geo=US&q=PFOS%20%2B%20PFOA%20%2B%20PFAS (accessed June 10, 2019).

148. Google Trends Public Data, 2019. Keyword search for "1,4-Dioxane". Available at: https://trends.google.com/trends/explore?date=all&geo=US&q=1,4-Dioxane (accessed June 10, 2019).

149. US National Library of Medicine National Institutes of Health, 2019. Results of key word search for "1,4-Dioxane". Available at: www.ncbi.nlm.nih.gov/pubmed/?term=1%2C4-Dioxane (accessed June 10, 2019).

150. US National Library of Medicine National Institutes of Health, 2019. Results of key word search for "PFOS". Available at: www.ncbi.nlm.nih.gov/pubmed/?term=PFOS (accessed June 10, 2019).

151. US EPA, 2015. Letter to Mayor David B. Borge regarding PFOA contamination found in the Hoosick Falls public water supply. Available at: www.epa.gov/sites/production/files/2015-12/documents/hoosickfallsmayorpfoa.pdf (accessed June, 2019).

152. US EPA, 2016. Drinking water health advisory for perfluorooctanoic acid (PFOA). May 2016. Available at: www.epa.gov/sites/production/files/2016-05/documents/pfoa_health_advisory_final_508.pdf (accessed June, 2019).

153. Agency for Toxic Substances and Disease Registry (ATSDR), 2018. Toxicological profile for perfluoralkyls: Draft for public comment, June 2018.

154. Chemical Watch, 2018. US NGOs press for release of PFAS tox profile. Available at: https://chemicalwatch.com/67587/ (accessed June, 2019).

155. Zitko, V., & Hanlon, M., 1991. Another source of pollution by plastics: Skin cleaners with plastic scrubbers. *Marine Pollution Bulletin*, 22(1), pp. 41–42.

156. Eriksen, M., Mason, S., Wilson, S., Box, C., Zellers, A., Edwards, W., Farley, H., & Amato, S. 2013. Microplastic pollution in the surface waters of the Laurentian Great Lakes. *Marine Pollution Bulletin*, 77(1), pp. 177–182.

157. De Graaf, J., Kaumanns, J., Koning, T., Meyberg, M., Rettinger, K., Schlatter, H., & Thomas, J., 2015. Use of micro-plastic beads in cosmetic products in Europe and their estimated emissions to the North Sea environment. *SOFW Journal*, 141 (4), pp. 40–46.

158. Boucher, J., & Friot, D. 2017. *Primary Microplastics in the Oceans: A Global Evaluation of Sources*. Gland, Switzerland: IUCN. 43pp.

159. Google Trends Public Data, 2019. Keyword search for "arctic pollution". Available at: https://trends.google.com/trends/explore?date=all&geo=US&q=arctic%20pollution (accessed June 10, 2019).

160. Google Trends Public Data, 2019. Keyword search for "microbeads". Available at: https://trends.google.com/trends/explore?date=all&geo=US&q=microbeads (accessed June 10, 2019).

161. US National Library of Medicine National Institutes of Health. 2019. Results of key word search for "arctic pollution". Available at: www.ncbi.nlm.nih.gov/pubmed/?term=arctic+pollution (accessed June 10, 2019).

162. US National Library of Medicine National Institutes of Health. 2019. Results of key word search for "microbeads". Available at: www.ncbi.nlm.nih.gov/pubmed/?term=microbeads (accessed June 10, 2019)

4 Conclusion
Managing Risks

This chapter describes a paradigm for anticipating emerging contaminants, drawing on information about the manufacture, use, and release of chemicals; developments in toxicology and ecotoxicology and the ways in which those developments are refracted through environmental regulations; and the human factors that can cause outrage to ignite or merely smolder. This eclectic analysis produces some ambiguity and contradiction, to be sure, but also provides a logical means to assess the likelihood that a contaminant will newly emerge as a concern. That understanding can support decisions to mitigate risk to human health and the environment, and ultimately to business concerns.

The Johari widow offers one way to view the analyses presented in this book and summarized below. Created in 1955 by Joseph Luft and Harrington Ingham [1] and then adapted for use in cost estimating and engineering (and made famous by former US Defense Secretary Donald Rumsfeld), this paradigm frames knowledge as follows:

- "Known knowns" are factors consciously understood. In the context of emerging contaminants, data from the Unregulated Contaminant Monitoring Rule (UCMR) program are known knowns.
- "Known unknowns" are variables one can identify but not quantify. Whether or not a chemical might be a carcinogen could be, for example, a known unknown until scientists perform conclusive tests. Public reaction may be a known unknown, though with the advent of predictive modeling based upon big data that may change.
- "Unknown unknowns" are blind spots. We simply cannot anticipate these factors based on past experience. One can imagine that the occurrence of persistent organic pollutants in the Arctic was for many years an unknown unknown, likely outside the realm even of conjecture.

The sections below describe known knowns and known unknowns and discuss how to use such information to anticipate the emergence of a contaminant. We must always, of course, be wary of the unknown unknowns that can upend thinking.

4.1 RECAPITULATION

The emergence of new concerns about a contaminant often follows a familiar arc. An issue can begin to emerge with a "fringe" finding of hazard or risk. If compelling enough, such a finding can spur academic research and may catch the attention of interest groups. Sometimes an issue simply smolders. Other issues may ignite, increasing attention from the media and the public. Then politicians feel compelled to take action, often legislating contaminants as a result.

A contaminant sometimes emerges as an issue due to new recognition of the hazard it presents. Our ability to understand hazards as a basis for identifying contaminants of concern has evolved slowly. The earliest efforts documented in this book date back more than a century. But much of the work in the United States occurred within a regulatory context between 1970 and the mid-1990s. The contaminants identified in those efforts and the characterization of hazards as reflected in regulatory decision-making have not markedly changed, although new insights into toxicity and ecotoxicity have changed the understanding of chemical hazards over time. As described in this book, the number of test methods for determining hazard have grown since the 1970s, as has have the nuances of interpreting test data. Regulators increasingly focus on effects that were only poorly characterized in the past, such as the potential for endocrine disruption or neurotoxicity, and apply those understandings to sensitive subpopulations. Further, re-interpretation of data using contemporary insights into the mechanisms of toxicity or effects on an ecosystem can change the understanding of dose–response and therefore the way in which we understand the hazards of a contaminant. Those new interpretations can change the levels thought of as "safe" for existing contaminants, or cause a chemical to emerge as a new concern.

The evolution of the Toxic Substances Control Act (TSCA) since 2016 and Registration, Evaluation, Authorisation and Restriction of Chemicals (REACH) since 2006 is changing the availability of data and driving the assessment of risks. Compliance efforts under REACH have produced an unprecedented amount of publicly available data on physical/chemical properties, environmental fate, and toxicity/ecotoxicity of chemicals. REACH dossiers for approximately 2,800 chemicals provide foundational information on physical/chemical properties, environmental fate, and toxicity/ecotoxicity. Regulatory submittals for nearly 4,700 substances manufactured at higher volumes provide even richer data sets that characterize additional forms of toxicity and the effects on additional species in the ecosystem. These data have become available just as amendments to TSCA require the US EPA to assess the risks of exposure to priority chemicals in commerce using the best available science. Those assessments, given the evolution of the science of toxicology within the last two decades and the increased availability of published data, will change the understanding of hazard for many chemical compounds. The result is likely to be changes in cleanup levels for some contaminants and the emergence of "new" contaminants.

The investigation and remediation of most sites focuses on fewer than 1,000 compounds. Regulators often base decisions on decades-old hazard characterizations of just 568 chemicals in the Integrated Risk Information System (IRIS) database. In contrast some 40,000 chemicals are manufactured in or imported into the United States, over 4,300 at more than 1 million pounds (approximately 450,000 kilograms) per year. Many of those high production volume chemicals present low hazard or exposure. However, some do present specific hazards:

- 50 High Production Volume (HPV) substances are considered to be persistent, bioaccumulative and toxic (PBT); of those, four are not on the Comprehensive Environmental Response, Compensation, and Liability Act (CERCLA) list of hazardous substances.
- 262 HPV chemicals are on the persistent, mobile organic compounds (PMOC) list; of those, 34 are not on the CERCLA list of hazardous substances.

Just because a chemical is produced in high volume and conveys a particular hazard does not mean that it will emerge as a contaminant of concern. To do so, it must be released into the environment, and then identified in a monitoring program using a laboratory method specifically designed to detect that compound at a concentration that may cause real or perceived harm. As a result, crucial to the identification of an emerging contaminant is the availability of analytical methods to detect the substance in environmental samples.

Several lines of evidence may indicate the potential for exposure. This book describes some of the efforts by governmental and nongovernmental agencies to collect data from industrial emissions or measure the concentrations of specific contaminants in surface water or groundwater, drinking water supplies, or biotic samples. Each of those programs focuses on a specific list of contaminants that have already emerged, in some sense. These lists of analytes include a fraction of the chemicals in commerce, generally those for which analytical methods are available and hazard data have been available to regulators. Consider two monitoring programs and the state of their evolving analyte lists:

- Under the Toxic Release Inventory (TRI), US EPA compiles data on 695 chemicals listed under 33 chemical categories. This list does not include all of the chemicals of potential concern. For example, 45 chemicals manufactured or imported at high volume (HPV) and classified as PBT are not tracked under TRI.
- The Contaminant Candidate List monitored in public water supplies under the UCMR program has varied depending on the phase of the program, from 50 compounds in UCMR1 to approximately 100 in UCMR3 and UCMR4.

The data resulting from such monitoring programs may exonerate some chemical substances, as illustrated in Chapter 3 by the data for some UCMR1 chemicals, or identify others as being of concern.

The final piece of the puzzle of why contaminants emerge as concerns is public reaction. Outrage drives perceptions of risk and creates demand for regulatory solutions. As contagious thoughts spread through social media, calls for action can intensify quickly and lead to new regulations or policies.

It is tempting to include in this summary lists of the chemical compounds identified through the analyses described in this book, and then to offer grand predictions about the emergence of such chemicals as contaminants of concern. But that would be irresponsible. While the analyses herein point to why and where contaminants emerge, the status of any one chemical compound depends on myriad factors. Lists, taken out of context, could incite a witch hunt instead of the reasoned scientific inquiry this topic deserves.

Instead of watch lists, the authors have developed a logical process for anticipating the emergence of a contaminant. The next section describes that process.

4.2 ANTICIPATING AND MANAGING DEVELOPMENTS

The case studies and data analyses described in this book tell us the signs to look for that may indicate the emergence of new contamination concerns. Early identification of a potential concern allows time to anticipate and manage developments. From that perspective, the sections below offer diagnostic factors regarding the potential for a contaminant to emerge and a logical process by which to evaluate a portfolio of sites to assess potential business risks.

4.2.1 POTENTIAL FOR CONTAMINANT TO EMERGE

A look backward at contaminants that have emerged as issues in the past suggests the stages of emergence, relative timelines, and relevant factors shown in Table 4.1. Each of these stages is described below.

4.2.1.1 Stage 1: Early Warning Signs

The first indications that a contaminant may emerge as a concern often happen, anecdotally, 2 to 20 years before the contaminant becomes known to present a possible risk. As illustrated by case studies discussed in this book, early warning signs include:

- Initial publications showing potential exposure or harm. Sometimes those findings emerge peripherally from a scientist's study of another compound. One can access this information based on conference presentations and indices of publications such as PubMed.
- The chemical has properties which suggest that it may meet the criteria for PBT, SVHC (Substances of Very High Concern), or PMOC. This factor may be difficult to predict if data are lacking. Quantitative Structure Activity Relationship (QSAR) tools can enable one to predict the effects of exposure, with varying degrees of accuracy, based on physicochemical data.
- The chemical has had wide dispersive use or release and toxicity has not been well understood or at least poorly characterized to date. Data emerging under REACH or QSAR analysis may be useful in screening

TABLE 4.1

Stages of Emergence

Stage	Potential Time to Emergence (years)*	Relevant Factors		
		Exposure	Hazard	Outrage
1 Early Warning Signs	2–20+	Dispersive use, particularly with uncertain hazard; Being considered for inclusion in a monitoring program	Early scientific publications; Potentially PBT, SVHC, or PMOC	Often dormant in early stages; Consider potential triggering factors for outrage
2 Signs of Possible Emergence	1–5+	Inclusion in a monitoring program; Momentum building in scientific publications	New data or reassessment planned under regulatory program	Visible signs of public interest and issue shows signs of contagiousness
3 Emerging Contaminant	0 – ~3	Monitoring program provides evidence of widespread contamination	Toxicological data show previously unknown hazards or more severe hazards	Inflection point reached in growth of public interest, potentially based on catastrophic event; Legislators begin to engage

*Timescales are suggestive, not absolute. Some chemicals never emerge as full-blown issues.

such compounds to see if they could potentially present a significant hazard or risk.

- The chemical is being considered for inclusion in a monitoring program such as UCMR, AMAP (the Arctic Monitoring and Assessment Programme), or another program, but is not yet formally included. This occurs when regulators or NGOs have identified some initial cause for concern, which they typically document in reports describing pending decisions. In some cases, proposals are open for public comment. That can allow the opportunity to contribute to science-based decision-making.
- Preliminary evidence suggests that a chemical or its use may trigger one of the factors that contribute to public outrage, especially if linked to a memorable event. This is enhanced if the idea of fairness comes into play where certain populations over others are "subjected" to the chemical.

At this stage it may be difficult to predict the eventual outcome with any certainty. There may be few known knowns; identifying the known unknowns and determining which to resolve can be crucial.

The ability to characterize the potential for exposure may be limited at this stage by the lack of validated analytical methods to enable scientists to detect a compound in environmental or biotic samples. The development of reliable analytical methods may be necessary before a suspected contaminant can advance to Stage 2.

Initial data showing potential exposure or harm may not receive much attention until substantial data accumulate, a regulatory change catalyzes a review, or public outcry begins. Thus, substances identified by their early warning signs may have a long latency period or may never emerge as contaminants of concern. But identifying a chemical through early warning signs offers the opportunity to amass scientific information, if appropriate, and to contribute to the public discussion, which will shape the understanding and treatment of the chemical.

4.2.1.2 Stage 2: Signs of Possible Emergence

Stage 2 represents the next progression in awareness. Data are beginning to build, and with those data the potential for a contaminant to emerge. Put another way, the known knowns are increasing though known unknowns certainly still exist. The outcome is not yet certain and often time and opportunity exist to influence the outcome. Stage 2 evidence, which may appear one to five years before a contaminant emerges, includes the following.

- The substance is included in study in a monitoring program such as UCMR or another large-scale program. This occurs when regulators or NGOs perceive enough of a potential threat that they are willing to invest in collecting data. At this stage, chemists may need to develop analytical methods to enable the detection of a compound in environmental or biotic samples; focused efforts to do so provide additional signs of possible emergence.
- Increasing numbers of scientific publications appear. Preliminary concerns over a chemical expressed in initial publications often generate more research and resulting publications. As illustrated for several compounds in this book, tracking the number of publications through databases like PubMed can provide an indication of whether or not momentum is building.
- New toxicological data have been developed or regulators have identified a substance for reassessment. New data published under REACH or in the scientific literature can upend previous hazard assessments embodied in regulatory frameworks such as IRIS. Further, designation of a substance for evaluation under programs such as TSCA can upend the previous understanding of hazard and risk.
- Public awareness is increasing. In this book, we've illustrated increasing public awareness of several contaminants by Google Tracking.

Software also allows one to track other measures of public interest, including blogs, tweets, and mentions of a substance in conventional publications.

Such evidence must be weighed carefully. After all, new data may show that a chemical is not widespread in the environment or may be less toxic than feared. Public interest might be neutral, or at least never ignite. But chemicals that trigger one of these factors bear watching. In the early stages of emergence, there is still time to contribute to science-based discussion and interpretation of data.

4.2.1.3   Stage 3: Emerging Contaminant
Contaminants emerge when:

- Monitoring programs such as UCMR provide evidence of contamination, and/or
- Toxicological data show previously unknown hazards or more severe hazards, which draw attention from legislators or regulators, and/or
- Public outrage ignites and spreads, sometimes with extraordinary speed.

While the business consequences may occur immediately, the full implications of emergence may take years to crystallize. It takes time for regulators to set cleanup levels, or for activists to organize bans. That may allow time for defensive actions but once a contaminant has fully emerged there is often little opportunity to influence the outcome of the process.

4.2.2   EVALUATING PORTFOLIO OF SITES AND DEVELOPING A RESPONSE

A company rarely evaluates a single chemical and its emergence as a concern as a hypothetical exercise. Instead, these efforts are often undertaken because they may impact business operations or they may affect current or future hazardous waste site cleanup. New information on the hazard and risks of a substance has implications for: defining or re-defining the boundaries of contamination at a site; the need for or changes in the type of remediation; and the costs that must be expended as a result. Described below is a step-by-step method to assess the overall impact of the emergence of a contaminant, by examining the known knowns and the known unknowns. Depending upon the circumstances these steps may be taken in different order and may be iterative.

Step 1: Begin by assessing potential for exposure. A key initial step is understanding if and how your chemical may be released into the environment. To gather this information, identify current and historical manufacture, storage, or use of the chemical, on its own or in mixtures. Consider whether it may be present as an impurity, by-product, or additive. The contaminant 1,4-dioxane, for example, emerged as a result of releases of certain chlorinated solvents. For many years, approximately 90% of 1,4-dioxane was used as a stabilizer for 1,1,1-trichloroethane. In addition, some paint strippers, antifreeze, and aircraft de-icing fluids contain 1,4-dioxane as a by-product [2]. Finally, data indicate that 1,4-dioxane is formed in many personal care products as a result of the ethoxylation process. All such processes and exposure routes should be identified in the initial steps.

Assess the potential for release, controlled or uncontrolled. Old clichés suggest that we identify contamination after finding a cache of rusting drums oozing foul liquids. But releases may result, as described in this book, from past accidents, industrial release (sometimes under permit), through outdoor use, use by the public, or as an impurity in a chemical released by one of these means.

Consider, too, the form of the chemical, its solubility, volatility, and propensity to sorb to solids. Those properties indicate whether a contaminant may emerge in surface water or groundwater, sorb to soil or sediment, or be of concern with respect to air pollution or long-range transport by air.

Step 2: Consider risk factors related to exposure and hazard. In step 2, evaluate the magnitude of potential exposure to receptors and the likelihood of adverse health effects to either humans or the environment. Releases that may affect sensitive or visible receptors deserve particular attention. Consider whether releases occur at or could readily migrate to settings near drinking water sources, surface water (particularly fisheries or water bodies used by subsistence fishers), residences, agricultural areas, or sensitive habitats and whether those releases have been widespread or isolated. Releases in remote areas may present lesser risk; both risk and perception of risk may be higher in areas where children might be exposed or in other circumstances that could spark outrage.

If an emerging contaminant is an impurity, by-product, or additive in another product that may have been released to the environment, consider the relative potential to migrate through soil and water. PMT/PMOC compounds like methyl tert-butyl ether [3] and 1,4-dioxane, once discovered, expanded the boundaries of plumes; as illustrated by sites contaminated with 1,4-dioxane, emergence of a new contaminant at an existing site can require re-treatment of a groundwater plume after remediation has been completed.

Appraise the information about the chemical's hazards, including not only the type of hazard that may be presented to human health or environmental receptors, but the amount of available data and certainty or ambiguity in the characterization. Methods that allow for comparisons with similar chemicals may enable estimation of possible effects and dose–response, if reliable data are not yet available.

Assess, too, the ability to get meaningful media-specific data at relevant concentrations. Analytical methods may or may not exist that can detect the substance at levels that could be of concern for health effects or are in line with regulatory limits. Further, monitoring data may raise more questions than they answer if the analytical methods are not accurate or precise, or hazards are not yet quantified.

Gauge public response and evaluate whether or how it is growing. Don't dismiss early warning signs of outrage as technically unfounded. Outrage can drive action, and if the outrage becomes contagious, the emergence of a chemical is likely imminent. Consider data gathering and public communications appropriate to establish an information base before outrage takes root and spreads.

Step 3: Assess and Plan. Finally, based on the available data, quantify the potential exposures, hazards, and resulting impacts and evaluate the likelihood that the chemical could evolve into an emerging contaminant.

Conclusions should be drawn for the chemical overall as well as evaluated at any relevant hazardous waste sites. If necessary, scale findings to a portfolio of sites. Assess the certainty of known knowns; consider which of the known unknowns should be resolved quickly; and remain aware that unknown unknowns can affect the outcome of any strategy.

These analyses support the development of a response strategy aligned in complexity with the scale of the impact and its likelihood. Appropriate responses may include:

- Monitoring the situation on both a technical and social basis
- Participating in trade groups' advocacy efforts
- Collection of data or support for research
- Revisions to site remediation plans and/or reserve estimates, if the impact is likely or imminent
- Communications internally and externally to support an appropriate response.

4.3 EPILOGUE

The analyses in this book show the potential for new contaminants to emerge from the gaps in our previous knowledge. Decades of experience in site investigation and remediation provide guideposts to how and why contaminants emerge. Following these guideposts, we can methodically assess the potential for release and exposure and assess the possible hazards, conscious of what we know and what we do not, to build a reasoned understanding and response. Our responses must also reflect an understanding of the forces that trigger outrage and cause that outrage to spread. However understandable that outrage may be, it can move decision-making further away from a science-based response and the prudent use of resources. Perhaps by better understanding the forces that drive the emergence of a new chemical, and working with knowledge of how outrage grows and spreads, we can better prepare and respond to issues as they arise and before they become a public emergency.

We offer these tools with caution and with hope. No methods predict the future perfectly. But the authors hope that the analyses described in this book contribute to the protection of our environment. Perhaps the best way to close this assessment of emerging contaminants and their management is with a quotation attributed to Albert Einstein:

> Learn from yesterday, live for today, hope for tomorrow. The important thing is not to stop questioning.

REFERENCES

1. Luft, J., & Ingham, H., 1955. The Johari window, a graphic model of interpersonal awareness. Proceedings of the western training laboratory in group development, 246.
2. US EPA, 2017. Technical Fact Sheet – 1,4-Dioxane. November 2017. Available at: www.epa.gov/sites/production/files/2014-03/documents/ffrro_factsheet_contaminant_14-dioxane_january2014_final.pdf (accessed June 12, 2019).
3. US EPA, 2008. Chapter 13 (MTBE) of Regulatory Determinations Support Document for Selected Contaminants from the Second Drinking Water Contaminant Candidate List (CCL 2), June 2008. Section 13.1. Available at: www.epa.gov/sites/production/files/2014-09/documents/chapter_13_mtbe.pdf (accessed June 11, 2019).

Index